$CO_{2.1}$

$CO_{2.1}$

Beyond the EU's Emissions Trading System

David Merlin-Jones

Civitas: Institute for the Study of Civil Society
London

First Published January 2012

55 Tufton Street
London SW1P 3QL

email: books@civitas.org.uk

ISBN 978-1-906837-34-1

Independence: Civitas: Institute for the Study of Civil Society is a registered educational charity (No. 1085494) and a company limited by guarantee (No. 04023541). Civitas is financed from a variety of private sources to avoid over-reliance on any single or small group of donors.

All publications are independently refereed. All the Institute's publications seek to further its objective of promoting the advancement of learning. The views expressed are those of the authors, not of the Institute.

Typeset by
Kevin Dodd

Printed in Great Britain by
Disc To Print
London

Should you find yourself in a chronically leaking boat, energy devoted to changing vessels is likely to be more productive than energy devoted to patching leaks.

Warren Buffet

A man generally has two reasons for doing a thing:
one that sounds good, and a real one.

J.P. Morgan

Contents

Figures and tables

About the author

David Merlin-Jones is Director of the Wealth of Nations Project. He is a graduate from Exeter College, University of Oxford where he studied modern history. He joined Civitas in 2010 as a Research Fellow, focusing on economic issues and, in particular, British manufacturing and energy. He became Director of the Wealth of Nations Project in November 2011. He has previously authored *Chain Reactions: How the chemical industry can shrink our carbon footprint* (2011) and *The Big Problem for Small Business: The Case for an Industry Bank* (2011) and other Civitas reports including *Rock Solid? An investigation into the British cement industry* (2010) and *Economic Growth – could the Government do more?* (2011) (with David Green).

Acknowledgments

The author would like to thank Nigel Williams for his assistance in constructing this report, together with Peter Gambrill. Special thanks also go to Alexander Yudin of St. Peter's College, Oxford, Damien Morris of Sandbag and Ruth Lea for their comments. All conclusions, omissions and errors are solely the responsibility of the author.

List of Acronyms

BIS	Department for Business, Innovation & Skills
BRIC	Brazil, Russia, India and China
CCL	Climate Change Levy
CCS	Carbon Capture and Storage
CDM	Clean Development Mechanism
CER	Certified Emission Reduction
CESP	Community Energy Saving Programme
CPF	Carbon price floor
CRC	Carbon Reduction Commitment
DECC	Department for Energy and Climate Change
EAC	Environmental Audit Committee
EC	European Commission
EED	Energy Efficiency Directive
ERU	Emission Reduction Unit
EUA	EU emission allowance
EU ETS	European Union Emissions Trading System/Scheme
GHG	Greenhouse gas
GIB	Green investment bank
HFC-23	Trifluoromethane
IED	Industrial Emissions Directive
JI	Joint Implementation
LCE	Low-carbon economy
LCLC	Low-cost, low-carbon

LCPD	Large Combustion Plant Directive
MF	Marguerite Fund
NAP	National Allocation Plan
N2O	Nitrous oxide
Ofgem	Office of Gas and Electricity Markets
R&D	Research and development
RED	Renewable Energy Directive
RHI	Renewable Heat Incentive
RO	Renewables Obligation
tCO_2e	Gas quantity equivalent to a tonne of carbon dioxide
$ktCO_2e$	Kilo (1,000) tonnes of carbon dioxide equivalent
$MtCO_2e$	Mega (1,000,000) tonnes of carbon dioxide equivalent

List of Energy Units

Joule: An SI unit of energy, equal to the energy dissipated by an electrical current of one ampere driven by one volt for one second.

Kilojoule (kJ) = 10^3 joules
Megajoule (MJ) = 10^6 joules
Gigajoule (GJ) = 10^9 joules
Terajoule (TJ) = 10^{12} joules

Watt (W): A unit of electrical power, the conventional unit to measure a rate of flow of energy. One watt amounts to one joule per second.

Kilowatt (kW) = 10^3 watts
Megawatt (MW) = 10^6 watts
Gigawatt (GW) = 10^9 watts
Terawatt (TW) = 10^{12} watts

Watt hour (Wh): The watt hour is a measure of work, the watt is a measure of power. The amount of wattage times the amount of time is the amount of work done. It is not used in the International System of Units (SI). The SI unit of energy is the joule (J), equal to one watt second. The kilowatt hour is commonly used though, especially for measuring electric energy. One watt hour is equivalent to 3,600 joules (there being 3,600 seconds in an hour).

Kilowatt hour (kWh) = 10^3 watt hours
Megawatt hour (MWh) = 10^6 watt hours
Gigawatt hour (GWh) = 10^9 watt hours
Terawatt hour (TWh) = 10^{12} watt hours

Megawatt and megawatt hour (MWh): A 1MW (megawatt) power-generating unit running for one hour produces one megawatt hour of electrical energy. MWe is used to emphasise when electricity is being measured and MWt is used when heat ('thermal') is being measured.

Therm: A unit of measurement of energy.

Reproduced by kind permission of Ruth Lea from Lea, R., & Nicholson, J., *British Energy Policy and the Threat to Manufacturing Industry*, Civitas, London, July 2010, p. 26.

Foreword

Prior to the establishment of the EU's Emissions Trading System (EU ETS) in 2005, a report of the Royal Society noted that:

> The basic issue is the current, largely cost-free emission of greenhouse gases into the atmosphere. A first step in reducing these emissions must be to use less energy. The second is to use energy sources that do not involve greenhouse gas emissions. The third is to ensure that the act of emitting greenhouse gases into the atmosphere ceases to be cost-free. These three broad approaches are intertwined.[1]

The reduction of emissions, notably carbon dioxide (CO_2), can be achieved by governments by advocacy, by regulation and by the application of economic instruments. The latter option, placing primary emphasis on the use of economic instruments, is generally held to provide the most cost-effective route where two generic types of economic instrument are possible. The choice lies between a carbon tax imposed on all CO_2 emissions or the allocation of tradable permits. Applying a penalty directly to the parameter that one wishes to reduce using the simple concept of direct taxation of carbon emission commends itself. To secure an emission reduction quantity target with a tax (a price), however, there must be fairly certain knowledge about the relevant price elasticities of demand for energy, transport, etc. The EU therefore chose the alternative: the tradable permit scheme, the EU ETS. Whereas a tax sets a *price* (the tax) and leaves the polluter to adjust the *quantity* (the level of emissions), a tradable permit system sets a *quantity* (a quota of emission permits), and the *price* (the price of the permit) adjusts according to the resulting supply and demand for permits.

The deficiencies in the market that make emission of CO_2 cost-free should be corrected by the scheme. Further looked-for consequences would be to make fossil fuels more expensive for the consumer, thereby encouraging switching to lower carbon technologies and energy sources, and reducing consumption through conservation and efficiency. In addition, it should make renewables, nuclear and carbon sequestration more viable. At least, that was the theory we understood in 2002 at the commencement of the EU ETS scheme when it was believed to be able to secure the environmental targets in question.

Unfortunately, in practice, over the intervening years, the scheme's achievement has not matched the theory. It is evident that the EU ETS was not properly thought through when it was planned. This report provides a damning description of incompetence in the design and application of the economic instruments in the EU ETS as it has evolved.

As described, the scheme has resulted at best in marginal carbon emission reductions and is beset by lobbying, corruption, huge profiteering, downright crime, carbon (and jobs) leakage through the emigration of manufacturing companies plus the creation of fuel poverty.

Nowadays the robbery of society through such a massive destruction of consumer wealth is simply explained away and justified by the phrase *market failure;* two hundred years ago even minor robberies were punishable by public execution! This state of affairs cannot be tolerated. Quite rightly this report concludes that the EU ETS should be replaced with other arrangements for emission reductions. As noted, however, there are too many vested interests involved at all levels, financially, intellectually and politically to justify expectations of changes in the near future.

In this respect, the American economist Thomas Sowell observed:

> Dangers to a society may be mortal without being immediate. One such danger is the prevailing social vision of our time—and the dogmatism with which the ideas, assumptions, and attitudes behind that vision are held.
>
> It is not that these views are especially evil or especially erroneous. Human beings have been making mistakes… as long as there have been human beings. The great catastrophes of history have usually involved much more than that. Typically, there has been an additional and crucial ingredient—some method by which feedback from reality has been prevented, so that a dangerous course of action could be blindly continued to a fatal conclusion… [2]

The EU ETS is but an ingredient of energy policy that has become part of a prevailing social vision of our time. It is a vision that is increasingly being transformed into a vast experiment for field-testing ideologically motivated, unilaterally promoted ideas for energy supply, use and now emission reduction. It is devoid of any consideration of costs to the consumer, efficiency or even feasibility of supply. The 'catastrophe of history' facing consumers will be unaffordable costs and power cuts.

The advocacy by green environmentalists, NGOs and politicians for renewables to provide 90-100 per cent of electricity supply is but one instance founded on a visionary dogmatism with complete ignorance of technical constraints. In the GB mainly island system, for example, no amount of wind energy available will remove the need for conventional generation capacity at almost the same level as at present in order to provide load matching via ramping services, frequency control, inertial stability and voltage control. With such high levels of both partly used conventional and renewable generation capacity, the costs to the consumer will be unaffordable; without such a level of conventional generation capacity being available as needed, power cuts are inevitable – the fatal conclusion.

As argued in this report and elsewhere, a fundamental purpose of any energy policy should be first and foremost to ensure the provision of reliable and reasonably priced energy supplies and, secondly, the reduction of emissions to a minimum in that priority; and not the reverse. The first feedback from reality that will be noticed, otherwise, is that without reliable and affordable electricity supplies a modern society, as we know it, will not function. No one should underestimate this threat to our physical welfare and economy.

This report argues, as have many other independent and neutral commentators of substance over the last decade, that the three distinct requirements of energy supply – reliability, affordability and minimum emissions – can be addressed to a considerable degree by a wholesale switch to nuclear power stations that do not emit carbon dioxide. This is not an acceptable technology to many; however, perhaps the best perspective is that recommended in the conclusions of the Swedish Technology Foresight Programme, in which the long-term future is seen as belonging to renewables, but the bridge to the future involves nuclear power.

The dilemma now is thus the choice between

- global warming
- nuclear power
- keeping poor people poor.

Professor Michael Laughton, FREng

Preface

The story of the European Union's Emissions Trading System (EU ETS) and all its various faults is not a new one. Many groups and reports have been discussing various aspects of its problems since the facts of Phase I's incompetence came to light. It is not new either to call for a carbon tax to replace it. This particular debate even precedes the ETS, when the options of cap-and-trade or flat-rate tax were the two choices open to the EU as the foundation stone of the European approach to dealing with carbon emissions. What is new in this report is the tying together of an EU ETS critique and a statistical assessment of a carbon tax to provide an alternative scheme to reduce carbon emissions, with new timescales and potentials and, most importantly, with a clear, feasible aim that is both economically and environmentally desirable: low-cost, low-carbon (LCLC) energy.

This report will avoid any discussion of issues that would detract or distract from the main argument. As such, there will be no debate over the nature of anthropogenic global warming or whether nuclear power is an ethical power source. Instead, this work will deal with the facts of the situation. The EU *has* pledged to reduce carbon emissions by 2050 by 80 per cent on 1990 levels and nuclear power *is* low-carbon. Moreover, the report will focus solely on the reduction of CO_2 emissions. Of course, there are other greenhouse gases which are all damaging to various degrees, but given that the EU ETS focuses on reducing CO_2, it makes sense to limit the discussion to this.

The EU ETS is not the cheapest way to reduce emissions, nor the method to reduce them sustainably. Arguments in favour of its retention that simply choose to avoid these facts and centre themselves on the premise that it's better than nothing are irrational. Why would it be better to accept lobbying, corruption, profiteering, the creation of fuel poverty and marginal carbon emission reductions? The unwillingness to question the environmental policy *status quo*, to acknowledge that industry and clean air are not mutually exclusive, and can co-exist to the benefit of all, has been the Achilles' Heel of the deep green movement, eroding its credibility and alienating the general populace.

David Merlin-Jones, Westminster, December 2011

Executive Summary

The EU's Emissions Trading System (EU ETS) is the flagship mechanism by which the EU hopes to reduce its carbon dioxide (CO_2) emissions via the principle of cap-and-trade. It has run from 2005 and will continue at least until 2020. However, it has failed to have the impact hoped for and is not the cheapest method by which to lower CO_2 emissions. It needs dismantling and replacing, while retaining the same aim of providing emission reductions at the lowest necessary cost.

Why it fails to reduce net global emissions

- There has been a huge over-allocation of credits via under-ambitious targets. The 2012 caps for 20 member states, including the UK, are higher than the measured emissions in 2005. Many companies will not need to make any reductions in their emissions until 2016-18 so even the tightening of the EU ETS in 2013 fails to have a real effect.
- The over-allocation of free credits is leading to huge windfall profits as companies pass through the non-existent credit cost to consumers or sell their credits if unrequired. The power sector alone is likely to have made €16-€50 million by passing on non-existent costs to consumers.
- The EU ETS actually risks raising global emissions. Companies whose competitive advantage has been undermined by the EU ETS emigrate to countries with slacker emission regulations and then the EU imports their products. In the UK this means that from 1990-2005, while production of carbon has fallen by 15 per cent, carbon *consumption* has actually gone up by around 19 per cent via imports.
- Whether the price of EU ETS credits rises or falls, emissions will not be lowered. A rise will result in carbon leakage and, if the price falls, it will be cheaper for companies simply to buy credits rather than install emission abatement equipment.
- The operation of the Clean Development Mechanism (CDM), a source of offsetting credits for the ETS, is racked with corruption and profiteering, involving billions of pounds. All five main project validating bodies failed UN accountability tests.
- The CDM's subsidies for emission reductions mean some emissions are being deliberately created to be destroyed and generate CDM credits. The gas HFC-23 generates 11,700 credits at €12 per tonne destroyed, but costs only €0.17/tCO_2e to destroy: a 7,000 per cent markup.

- Volatile price fluctuations and the overall short-term nature of the scheme leave investors with little ability to plan ahead with surety: a huge problem.
- Crime in the ETS is rife. 90 per cent of all market activity in 2009 was estimated to be criminal by the European Law Agency and carousel fraud alone is estimated to have cost the EU €5 billion in lost tax revenue
- Other countries are not following the EU's lead. Beyond the Kyoto Protocol, Europe is the only region implementing an ETS, risking serious competitive issues. Other countries have examined and rejected them or implemented very weak schemes.
- Aviation is supposed to be included in the EU ETS from 2012, but the full details of how this will be applied are not yet known and tickets for 2012 are already on sale.
- Attempting to force inclusion in the EU ETS on non-EU sectors, as with aviation, is highly likely to provoke trade wars. Chinese, American and Russian airlines look set to avoid using EU hubs to escape the charge, meaning countless lost business opportunities for Europe.

Why it won't work much better in the future

- The problems of the EU ETS cannot be solved without undermining its market-driven nature, the *raison d'être* of the scheme that caused it to be chosen over a carbon tax. The EU has to admit that this market-based solution to carbon emissions does not work as desired and replace it.
- By using multiple approaches to try to patch the ETS up, the overall goal of minimal economic cost is undermined while critical flaws go unnoticed and emissions continue to rise.
- Current unilateral policies tangential to the EU ETS, such as the UK's carbon price floor, disadvantage a country's industrial capability. Raising the cost of production without industrial emigration is only possible on a wider regional level (although the ideal would be on a global scale) such as across the EU.
- There are too many vested interests holding it back from its full potential. The carbon market is geared towards maximising profits rather than emission reductions. Even the World Bank has invested nearly $2 billion in deliberately polluting Chinese HFC-23 factories, and pressured the EU to let them continue receiving these credits.

- Governments are not hypothecating. The UK is also cynically collecting the proceeds of ETS auctions while refusing to promise to spend these on green projects: it is effectively just another tax and this undermines its credibility.

A better approach to reducing emissions: target the power sector

- If the EU truly wants emission reductions together with a strong economy, this cannot be rushed. It is a long-term challenge with long-term solutions.
- The best way to reduce emissions is to target energy generation. In 2010, energy supply accounted for 39 per cent of British CO_2 emissions, a figure roughly in line with EU energy output. However, existing policies towards low-carbon power generation have put renewables first, and alleviating climate change and emissions second.
- Whatever source of power is used, energy should be low-cost to ensure a thriving and competitive economy. This means there is a strong argument to develop more nuclear power, which is also low-carbon.
- It is a waste of resources to rely on the comparatively inefficient renewable energy sources of the present day and the Renewables Obligation and Renewable Energy Directives should be ignored. For the UK, the only target that truly matters is the long-term 80 per cent reduction in CO_2 on 1990 levels by 2050. Whether this is reached via renewables or not is irrelevant.

The new goal: low-cost, low-carbon (LCLC) energy

- To meet a theoretical global target of reducing emissions by 80 per cent by 2050, even if OECD nations produce near nil emissions, this will only allow others to emit 2-2.5 tonnes of CO_2 per capita. The average Western country currently emits 20 tonnes of CO_2 per capita.
- The true target of environmental policy should be to develop low-carbon power so the cost falls to the level of fossil-fuel energy *without* carbon taxes and then these power sources will sell themselves, without subsidies or penalties. Plentiful, cheap energy is in no way a bad thing, if it is low-carbon.

- Lowering the cost of low-carbon power will lead to its adoption worldwide. This will reduce global emissions in a manner that treaties and the CDM have so far failed to do. EU support for nascent LCLC will establish Europe as the principal global location for an advanced power industry.
- The widespread use of low-carbon power is not going to happen in the short term, regardless of the governmental policies trying to promote renewable power. If investment in low-cost, low-carbon (LCLC) power sources is maximised, they will be available for mass deployment around 2030.
- In the meantime, there are quick and easy ways to balance the need for cheap energy and ensure this is low-carbon. The government should promote the construction of more current generation nuclear power stations, or, if this is seen as politically undesirable, encourage power generators to switch from using coal as a fuel to gas.

The method: replace the EU ETS and all other green costs with an EU-wide carbon tax to fund LCLC development

- The EU should dispense with the EU ETS, and all other Directives aimed at reducing emissions via levies and taxes, and all unilateral actions should also be halted. These should then be replaced with a flat-rate carbon tax charging '£X' per tonne of CO_2 emitted, with the tax rate rising or falling according to scientific advances in the field of climate science. In this report, £30/tCO_2 has been used as an illustration, because this is the level set by the carbon price floor for 2020. More detailed research into the economics of such a tax might suggest that a lower level would be preferable.
- Initially, for the sake of simplicity, the carbon tax should cover the same installations currently included in the EU ETS, given that they are the largest emitters. This could be extended to other sectors or domestic gas rates if felt desirable. Domestic electricity would receive the tax passed through from generators.
- This will generate the revenues needed to fund intensive LCLC R&D and alleviate resulting fuel poverty, neither of which are currently funded by EU ETS revenue.
- At £30/tCO_2, £9 billion would be raised, enough to eradicate fuel poverty in a year and deliver £1 billion to be spent on LCLC research. Thereafter, the full sum would be available for LCLC investment. This is substantially less than current policies will cost and will deliver far greater revenues and greater long-term reductions.

- If the EU is unwilling to dispense with the EU ETS, the UK should do so anyway, and adopt the flat-rate carbon tax for LCLC research by itself. Britain could do this through repatriating the relevant powers; it does not necessarily have to leave the EU.
- If the British Government wishes to back up its claim to 'lead on climate change', then replacing ineffective mechanisms with a viable one is necessary. To fail to do so will undermine its environmental mandate. With the UK already leading in the field of marine energy generation, and a wider tradition of strong research, it is ideally suited to deliver LCLC energy for the world.

Introduction

The European Union's Emissions Trading System (EU ETS) has been the flagship European response to global warming.[3] It has been designed to limit the emission of carbon dioxide (CO_2) from the largest European CO_2 emitters, in order to reduce the level at which it is building up in the atmosphere. Crucially, the ETS was set up with the professed aim of helping participating countries 'limit or reduce greenhouse gas emissions (GHGs) in a cost-effective way' and the reputation of the scheme therefore relies on the EU ETS being the cheapest way to deliver these reductions.[4] Disappointingly, this appears untrue. It is an expensive scheme, and not just in financial terms. It has caused costs to rise for companies and consumers, and there is also an environmental opportunity cost given its numerous flaws that make it less efficient at reducing GHGs than other methods: because of the EU ETS, emission levels are not being reduced as rapidly or as cheaply as they should be.

This raises the question of what Europe should be doing. Does the EU ETS need to be overhauled, but retained roughly in the form in which it now exists? Yes, say many environmental campaigners, along with the EU itself. Their arguments are partially based on the fact that it *is* the only large-scale emission-reducing scheme attempted: better the devil you know. While it has its faults, so the defence goes, any weakness is an opportunity for constructive engagement to improve the scheme. Its potential is its greatest asset.

Alternatively, there are others calling for the ETS to be scrapped. These include the obvious climate-change sceptics, who see the project as a means to introduce new EU taxes by the backdoor (which is true in part), but also environmental extremists who feel the ETS is too much of a compromise, corrupted by capitalism and open to profiteering as well as lobbying (which is also partially true). Green issues produce strange bedfellows.

Thirdly, the most attractive option: the dismantling of the EU ETS and its replacement by something else. In this report, a flat-rate European-wide carbon tax is presented as the optimum solution, provided that all the proceeds are sensibly injected into the development of long-term low-carbon energy sources, with the simple aim of making these as competitive as low-cost fossil fuels.

This report is split into three sections, the first of which examines the state of the EU ETS and its major flaws. The second part examines the UK's responses to the failure of the EU

ETS. Finally, the third section outlines what the EU ETS should be replaced by, and argues for a paradigm shift in environmental thought, away from just wanting to curb emissions symptomatic of carbon-intensive societies, and towards developing the holy grail: low-cost, low-carbon energy.

How the EU's Emissions Trading System works

Superficially, the EU ETS appears a very simple scheme. It is uses the cap-and-trade principle, which means that the upper limit of European CO_2 emissions is set, but within that level participants are free to buy and sell permits as they see fit. The allowances can be provided in the form of specific EU emission allocations (EUAs) or international credits, the most prevalent of which is the Certified Emissions Reduction credit (CER) which is discussed in detail below (see p.17). All credits are worth the same value in terms of CO_2 reductions, that is, the reduction of one tonne of CO_2 equivalent (tCO_2e). The System does not cover just CO_2, but the wider Kyoto basket of six greenhouse gases. CO_2, despite being the least potent of the gases, made up 80 per cent of the EU's weighted GHG emissions in 2009 and is therefore the main GHG the EU ETS seeks to reduce.[5]

Companies acquire their credits by two means. Currently, the vast majority are allocated, free of charge, to EU ETS compliant installations. At the end of each year, installations have to declare how much CO_2 they have emitted, translate this into how much of their allowance they have used and surrender the corresponding number of credits. If some credits are left, these can be sold through the carbon markets to others who emitted more than their allowances permitted and need to make up the difference. Additionally a smaller but growing number of credits are auctioned off officially, allowing installations to top up their allowances further. This occurs within the 'carbon market', as it is known, and the fluctuating prices of these credits can be seen in Figure 0:1 (see p.xxv). In principle, all this rewards companies who reduce their GHG emissions and penalises heavy emitters.

The number of allowances received freely by EU ETS installations is currently decided at a national level. Under the first two phases of the scheme, each EU member state creates a national allocation plan (NAP) with their planned emission cap, which then has to be approved by the European Commission (EC). Once accepted, each country allocates their free allowances to its industries via 'grandfathering' which is allocation based on a company's previous emission levels. This means that the companies emitting the most CO_2 receive the most credits. From Phase III, the NAPs will be replaced by an

Figure 0:1
Average trading prices of carbon dioxide emissions

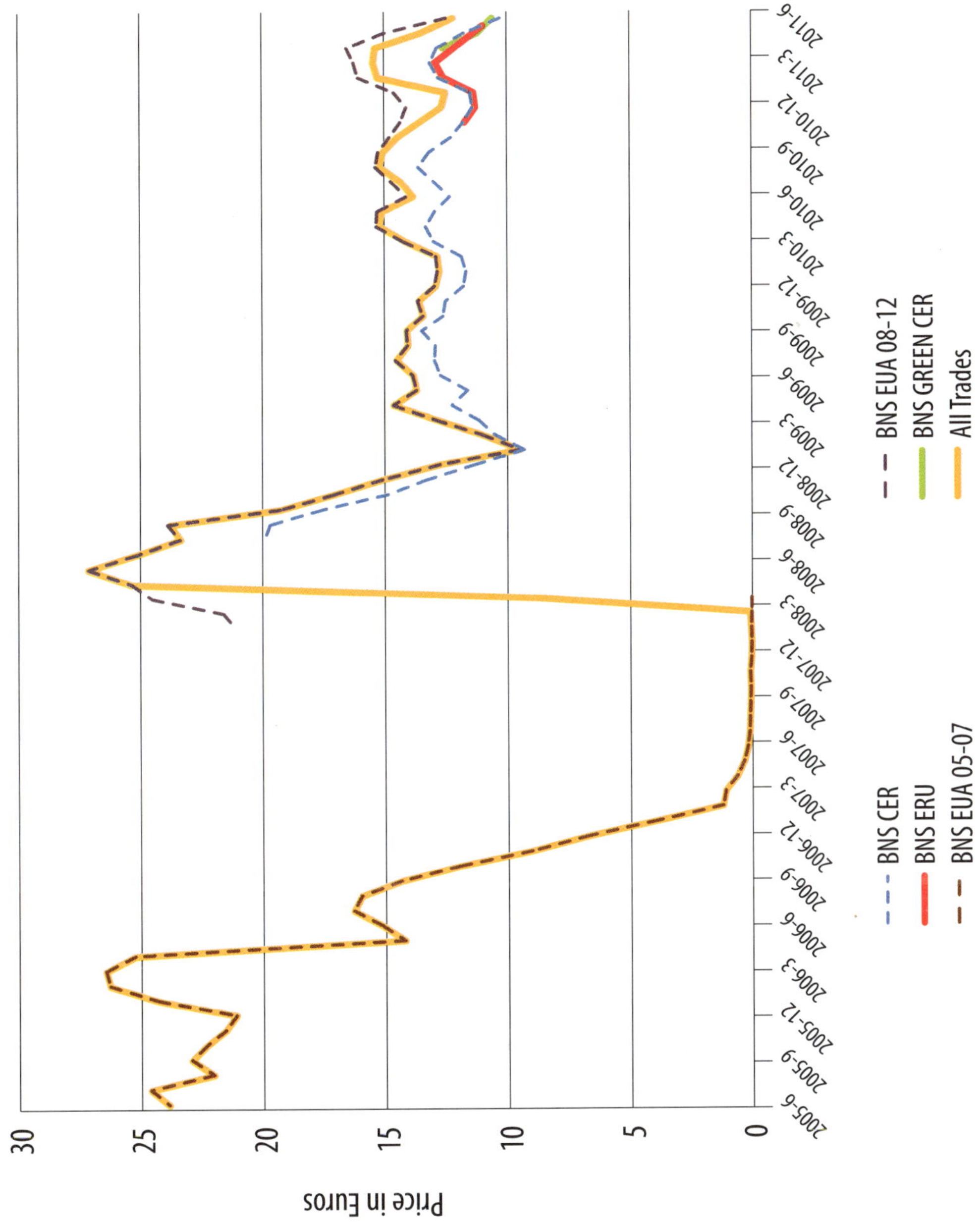

Source: BlueNext, Historical Transactions

EU-wide cap and grandfathering will be replaced by benchmarking, where the remaining free allocations will be allocated according to the best industry-wide emission levels. In principle then, the most environmentally–friendly companies will receive all the credits they need while the least environmentally-friendly will not and will have to pay for extra credits and/or reduce their emissions. These harmonised, pan-European rules are meant to ensure the system is fairer as industries, not countries, are judged.

The EU ETS has been running since 2005, and was initially set up as the EU's response to the Kyoto Protocol which obliged Europe to reduce its emissions in 2008-12 by eight per cent on 1990 levels. Now, the overall goal is to reduce emissions by 21 per cent on 2005 levels by 2020. There are three stages to the System, known as 'phases':

1. Phase I (2005-07): Complete

During the initial phase, the EU ETS included 12,000 installations and set a cap at 2,298.5 $MtCO_2e$. Those sectors covered included energy activities such as combustion installations with a thermal input greater than 20 MW (including power plants), oil refineries and coke ovens. Additionally, energy intensive industries were targeted, including the production and processing of ferrous metals; mineral industry sectors such as cement, glass and ceramic bricks; and pulp, paper and board activities.

During Phase I, European emissions rose 1.9 per cent, but did not actually exceed the cap in any year.[6] Over-allocation of credits caused the price to crash. Phase I credits could not be carried over to Phase II.

2. Phase II (2008-12): On-going

The second phase of the EU ETS set the new cap at 2080.93 $MtCO_2e$, a reduction of 9.5 per cent on the Phase I target.[7] CER credits were now introduced as well, as an alternative means for installations to obtain offset credits. This was part of a drive to link the EU scheme to wider global environmental initiatives via the Kyoto Protocol, and Europe has converted nearly half of its Kyoto Allowances into EUAs. During this period, the free allocation of EUAs continues, with auctioning only accounting for 7 per cent of activity.[8] From 1st January 2012, aviation will be included in the System (see p.44).

Figure 0:2 (see p.xxviii) displays the difference between actual emissions produced by countries, and the issued allowances to member states.

Figure 0:2
The surplus and deficit of EU ETS allocations by country through all trading periods

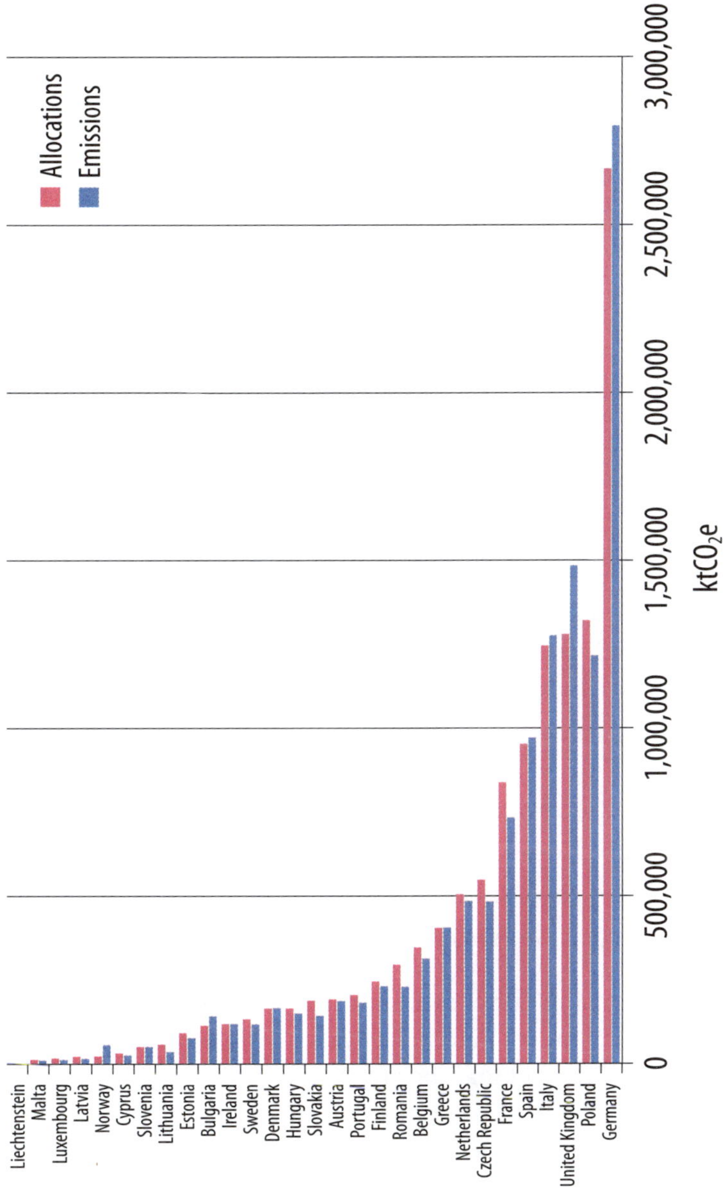

Source: European Environment Agency, EU ETS data viewer

3. *Phase III (2013-2020 (probably longer)): Future*

The final phase will increase dramatically the pace of the ETS to reduce emissions by 21 per cent on 2005 levels by 2020. Auctioning will become the norm and increase in a linear manner, up from four per cent of allocation at present to 50 per cent in 2013, 70 per cent by 2020 and 100 per cent by 2027. The power sector faces 100 per cent auctioning from Phase III's inception. In theory, this will end the generation of windfall profits occurring at present. A non-binding clause in the ETS legislation recommends that half the revenue from this auctioning should be spent on funding initiatives to curb climate-change both inside and outside the EU. 12 per cent of the auctioned allowances will be redistributed to poorer EU states 'in the interest of solidarity' and most of these will be new Eastern European members.[9]

The total cap will also be reduced year-on-year in a linear fashion. The initial cap will be the average allocations of the 2008-12 period, adjusted to reflect the widened scope and the exclusion of any small installations. From that number, a reduction of 1.74 per cent will take place annually and will come up for revision in 2025. The expectation is that this narrowing will raise the price of permits and long-term polluters will therefore be penalised. Allowances from Phase II will be transferable over to Phase III, a problem we shall come to soon. For a theoretical example of how a company is affected by the EU ETS, see Appendix 3 (p.127).

The ideal

Before discussing the problems associated with the EU ETS, it would be sensible to recount why it was chosen over a carbon tax in the first place. Back in July 2003, the Directive laying the foundation of the EU ETS was passed. The then Environment Commissioner Margot Wallström heralded this breakthrough thus:

> The agreement on this Directive signifies a breakthrough both for climate change and emissions trading. It means that the largest emissions trading scheme in the world to date will be a reality from 2005, and that the architecture foreseen under the Kyoto Protocol is coming to life. Companies across 25 countries must now start incorporating climate change into day-to-day commercial decisions, and begin assessing what innovative steps they can take to reduce emissions.[10]

Primarily, the choice of a cap-and-trade scheme was to avoid the uncertainty issues that would burden a carbon tax. A tax would have been trickier to get right: setting the correct level to internalise carbon costs without overly burdening industry was deemed too hard, especially since later scientific developments could reveal this cost to be too high or low in retrospect. Overall, it was thought that a properly constructed ETS would be most likely to deliver the best results. According to the Lord Stern, the head of the eponymous review that called for policies like an ETS, there are three key principles it would deliver:

> i. Effectiveness: a cap imposes an absolute limit on emissions, and therefore clarity on reductions.
>
> ii. Efficiency: competition and the market will seek out the cheapest ways of reducing emissions.
>
> iii. Equity: the structure of quotas, together with the exploitation of some low-cost emission reductions options in developing countries, can generate private-sector finance to developing countries to support low-carbon growth [...] These financial flows can provide part of the 'glue' for a global deal.[11]

The flexibility of the ETS was the dealmaker. It would allow companies to decide how best to reduce their emissions, be it a new plant or new processes. In addition, any policy has to be deemed acceptable to the voting public and the ETS was judged so, although the extent to which the public is actually aware of the scheme's existence is questionable. For those concerned by fuel poverty, it offered a way to reduce energy related emissions without burdening the ordinary bill payer. Instead, it would be big business and power companies that would pay, justifiably given that they produced the emissions in the first place. Others, worried that the climate change agenda would mean the state creeping into their private lives or interfering with their businesses, were supposed to have their minds put at rest. With a market-based initiative, governments would step back and passively monitor the scheme. Both sides of the political spectrum would be satisfied, provided, of course, they felt climate change was an issue that had to be dealt with in the first place. Finally, the international context appeared to favour the ETS as policymakers optimistically assumed it could eventually link up with other schemes around the world, ultimately to create a global emissions trading system.

As this report shows, the EU ETS has failed to live up to the overly ambitious dreams of its architects and it has proved very hard to translate the ideals of theory into a successful scheme in practice. It has turned into a lose-lose scheme: at low credit prices, installations prefer to offset their emissions instead of investing in expensive low-carbon technology, and as the price of credits rises, some installations simply emigrate to extra-EU countries to avoid the cap. The EU ETS has failed then, and will continue to fail us. We therefore need a new goal. As will be discussed, this goal should be the intensive development of low-cost, low-carbon energy, funded through private and public investment. While low-carbon, or more specifically, renewable energy is already a high government priority, their timeframes and narrow definitions are neither suitable nor achievable, expecting a huge development in a short space of time to 2020.

The nascence of low-carbon power *is* currently underway, but the most effective solutions are highly unlikely to become available pre-2020, in part due to the need to drive down the cost to equality with carbon-intensive fuels. Just as we should avoid a carbon technological lock-in, so too should we eschew an inferior and expensive low-carbon lock-in. In the meantime, it would be wisest to concentrate on minimising the emissions from our existing fossil-fuel technology, which can be best achieved through switching from coal to gas as the mainstay of fossil fuel plants. Given the already sizeable power plant closures known to be occurring in the near future as a result of obsolescence and the EU's Large Combustion Plant and Industrial Emissions Directives, this is the only way to keep the lights on.

PART ONE

Diagnosis: The Problems with the EU ETS

CHAPTER 1

The Dilemma of the Cap-and-trade Model

- The 2012 caps for 20 member states, including the UK, are still higher than the measured emissions in 2005.
- In 2010, 65 per cent of businesses were still receiving more credits than they needed and throughout Phase II credits worth 400 million tonnes were given out when businesses didn't need them.
- As a result of overly high caps, installations accounting for 50 per cent of EU emissions, will have no need to make any carbon cuts until 2016-18.
- The EU ETS does not provide long-term investor confidence, and little is known about how it will run post-2020, reducing enthusiasm for investment in carbon-reducing technology.
- In 2009, a quarter of the surplus credits were in the hands of just ten companies, and as of 2011, they have received credits to the value of €4.1 billion during Phase II, four times the EU environmental budget of the whole period.
- Power companies are passing on the 'cost' of their free allocations to consumers by raising energy bills, earning an effective windfall profit. By the end of Phase II, the eventual gains have been estimated at between €16 billion to €50 billion.
- Ultimately the scheme is not about paying to emit, but about reducing CO_2 output. However, the UK Government has already estimated Britain will be a net purchaser of 14-25 million permits, thereby ignoring the real *raison d'être* of emission reductions.

Over-allocation of permits

For many supporters of the EU ETS, the fact that it contains an explicit limit on the amount of CO_2 that can be emitted is its greatest strength, and the reason for its choice over a carbon tax. The number of allowances offered should be very carefully chosen, in order to provide scarcity and therefore a catalyst for a self-sustaining price. In theory, once this knowable cap is chosen, it can then be reduced over time, reducing everyone's emissions in the process and hopefully causing the cost to rise. However, this has simply not been the case, and the scheme's supposedly most promising aspect has become its greatest frustration.

Figure 1:1 (see p.6) shows the caps of the three phases of the EU ETS. During Phase I, installations were emitting GHGs at a level well within the specified cap, not because they were trying to reduce their emissions but because the cap was set at a level significantly higher than their actual emissions. By the end of the period in 2007, emissions covered by the ETS were 38 million tonnes higher than they had been at the beginning of the scheme in 2005, a rise of 8.3 per cent. There was a surplus of 44 million allowances after the first year of the scheme alone.[1] As this continued, the price of EUAs entirely collapsed in 2007 from a previous high of €27 to €0.10 for the last six months of the Phase, as seen in Figure 0:1 (see p.xxv).

Phase I was conceived as a test period. Three years of establishing whether trading mechanisms could work saw limits set at a level that was intended to cause very little industrial hardship. Allowances could not be carried over into the next phase, which contributed to the carbon price crash in 2007. With few installations needing to buy credits for their own immediate use and no prospect of selling them at the later date, sellers greatly outnumbered buyers and the price fell. Despite limited incentives, trading in the early years of the phase included an element of fuel-switching. This involves no more stringent economies than changing the order in which power stations are taken in or out of production as demand changes in daily and seasonal patterns. With a carbon price, it is possible that, as a result of fuel-switching, high-emitting plants that would have been in use remain idle.

In the third quarter of 2005, Delarue, Ellerman and D'haeseler estimated that abatement from fuel-switching reached 16 $MtCO_2e$ across the ETS, using a model calibrated to results from 2003 and 2004.[2] The earlier years were a time when coal use was lower than might have been predicted, for a number of possible reasons. A simpler model, without the calibration, suggested abatement for the two years of close to 100 $MtCO_2e$.

Table 1:1 A comparison of proposed and actual National Allocation Plans

Million tonnes of CO_2	2005 verified emissions	Proposed cap 2008-2012	Cap allowed 2008-12	Percentage cap allowed in relation to proposed
Austria	33.4	32.8	30.7	93.6
Belgium	55.58	63.3	58.5	92.4
Bulgaria	40.6	67.6	42.3	62.6
Cyprus	5.1	7.12	5.48	77.0
Czech Rep	82.5	101.9	86.8	85.2
Denmark	26.5	24.5	24.5	100.0
Estonia	12.62	24.38	12.72	52.2
Finland	33.1	39.6	37.6	94.9
France	131.3	132.8	132.8	100.0
Germany	474	482	453.1	94.0
Greece	71.3	75.5	69.1	91.5
Hungary	26	30.7	26.9	87.6
Ireland	22.4	22.6	22.3	98.7
Italy	225.5	209	195.8	93.7
Latvia	2.9	7.7	3.43	44.5
Lithuania	6.6	16.6	8.8	53.0
Luxembourg	2.6	3.95	2.5	63.3
Malta	1.98	2.96	2.1	70.9
Netherlands	80.35	90.4	85.8	94.9
Poland	203.1	284.6	208.5	73.3
Portugal	36.4	35.9	34.8	96.9
Romania	70.8	95.7	75.9	79.3
Slovakia	25.2	41.3	30.9	74.8
Slovenia	8.7	8.3	8.3	100.0
Spain	182.9	152.7	152.3	99.7
Sweden	19.3	25.2	22.8	90.5
UK	242.4	246.2	246.2	100.0
Total	**2122.16**	**2325.34**	**2080.93**	**89.5**

Source: Europa, Press Release IP/07/1614, 26 October 2007

Figure 1:1

EU ETS Phase I to III caps, allocations so far and verified emissions

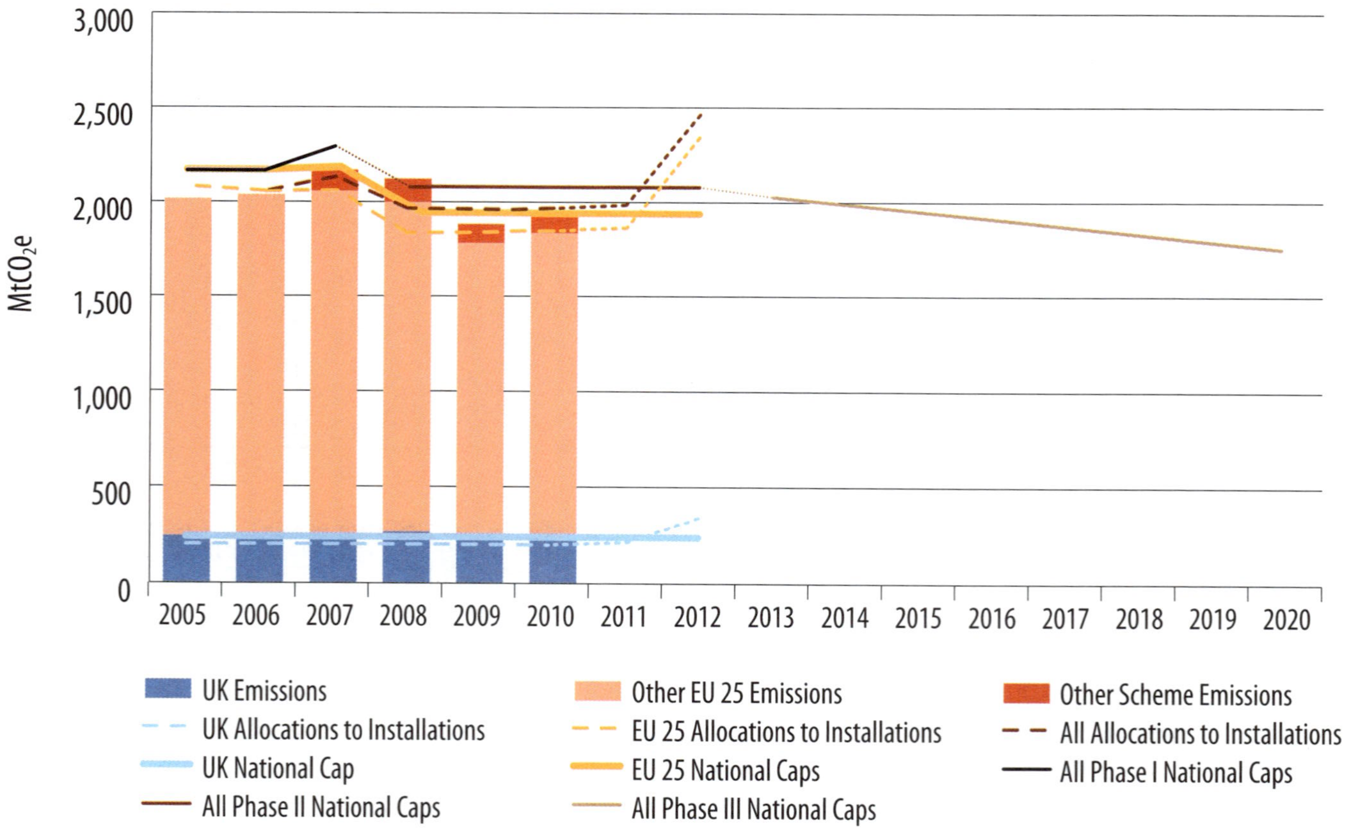

Source: See Appendix 4

In Phase II, over-allocation continued and, in 2010, it has been estimated that 65 per cent of installations were still being overprovided with permits. This has allowed them to continue to emit CO_2 as though the ETS did not exist and so it is unsurprising that, in 2011, 40 per cent of companies said it had not affected their 'business as usual' plans.[3] In part, this was due to the allocations being based on 'grandfathering' – i.e. permits being given out on the basis of previous emission levels – and the cap assuming that economic growth would occur during the period. The recession put a stop to this and resulted in a lower volume of production and, consequently, a smaller quantity of emissions. This was not reflected in the allowances, which were still being allocated at pre-recession levels. The outcome of this has been the give-away of 77 million surplus permits during 2008. For the whole of Phase II, it has been estimated that 400 million surplus permits will have been distributed.[4]

This glut will accelerate at the end of Phase II, as can be seen in Figure 1:1 (see p.6). The graph shows EU-25 limits separately, so as not to produce an obvious discontinuity. The Phase II limits do include some extra, energy-intensive installations, unrestricted in Phase I. These amount to 55 $MtCO_2e$, which is to say that an equal level of monitored emissions at the start of Phase II would represent a cut of 2.5 per cent.[5] So far, many countries, including the UK, have been handing out fewer Phase II credits than the cap allows per year. This means that there is a stockpile being produced, and if the intention is to release all these credits onto the market by the end of the second phase, there will be a steep rise in the credits allocated as this backlog of nearly 500 $MtCO_2e$ worth of credits swamps the market, the equivalent of over double the UK's verified emissions of 2010 (see Appendix 4 for details, p.130). If this occurs, this will not only crush the price of credits, as the huge flood means few will need to buy any for a while, and credits being two-a-penny, those that do will be able to do so cheaply as demand slackens. Those companies that decide to buy while the going's cheap or just hoard their own surpluses will be comfortably set to ignore the need to reduce their emissions for a good while during Phase III, despite the lowering cap.

The recession alone cannot be blamed for the over-allocation and the looming glut. The inability to take control of emission levels themselves has been due overall to the lacklustre targets set: the 2012 cap for 20 member states, including the UK, is still higher than the measured emissions in 2005.[6] These caps are equivalent to the number of credits to be given out and auctioned. Why the overly generous caps? Unsurprisingly and understandably, it was because everyone involved in the cap creation did not want to undermine their own competitive advantage. The ETS relied on industries submitting their own emissions statistics to their national governments, which then developed the NAPs. If this information is not accurate, then the NAP will be skewed accordingly. Obviously, it is in the industrial interest to overstate the volume of emissions in order to receive a greater number of allowances, and only the submitting companies really know how many they need. The magnifying effect could then be repeated at the national level, as countries pressed for the most generous caps in order to protect their own industries. Effectively, the EU ETS suffered from a 'race to the bottom', as member states had no wish to subject their industries to harsher measures than other EU members.

Table 1:1 (see p.5) shows the proposed NAPs for Phase II, along with those the European Commission (EC) agreed to. Only four countries, including the UK, had their proposed plans accepted without modification, while others had dramatic reductions made. Estonia's NAP was almost halved. Even after the modification of the total carbon budget for 2008-12, which was 9.5 per cent higher than the verified emissions of 2005, the final outcome was just 1.95 per cent lower: hardly the stuff of ambition.[7] This pathetic reduction was the final outcome of all the political pressure and lobbying directed towards the EC to minimise their obligations. Germany, which had its NAP reduced by just six per cent, even threatened to take legal action against the Commission. In the end Germany, backed down on the condition that it could use Clean Development Mechanism (CDM) and Joint Implementation (JI) credits to make up to 20 per cent of its allocation, as opposed to the 12 per cent originally specified in the NAP.[8] From 2008-10, it met 19.4 per cent of its target via this method.[9] As shall be seen later, this makes their emission targets considerably easier to reach and significantly reduces the burden on their industry.

Effectively, the EU ETS suffered from a 'race to the bottom', as member states had no wish to subject their industries to harsher measures than other EU members.

In addition, by setting underwhelming targets, the ability to judge the effectiveness of the ETS is reduced, undermining one of the founding principles of the scheme: its ability to control the level of emissions. If GHG levels are below the cap, which, as Figure 1:1 shows (see p.6), they have been in many years, this doesn't mean the System has been successful, as the allowed quantity could itself be too high, giving a wide margin for further emissions. This appears very likely, especially given that the initial caps of Phase I were greater than emissions produced. The original ethos was therefore one of limiting emission growth rather than actually reducing it. An ungenerous observer might also comment that high caps were good for the EU's publicity as well: if emissions remain within the cap, this suggests the ETS is working, justifying its original choice over a carbon tax.

As a consequence of the weak NAPs, the price of allowances on the carbon market has fallen considerably, as companies have had little need to buy extra ones so overall demand has been low. Clearly, this has significant repercussions for the principal goals of the EU ETS, as a low permit price fails to incentivise the installation of carbon-efficient equipment in industry. Instead, in the event that companies do need to do something to satisfy the ETS, the cheapest solution is to maintain current pollution levels and simply buy more allowances, abating the symptom, not the condition itself.

The over-allocation is a danger to the third phase of the ETS, given credits can be carried over from Phase II to III. The steadily growing surplus of credits means they have more leeway in Phase III and can continue to emit CO_2 with a barely modified 'business as usual' attitude. It has been estimated that when Phase II ends, credits worth 970 $MtCO_2e$ will be carried over, the equivalent of 40 per cent of the Phase II cap.[10] Left to itself, this will be resolved given enough time, as the linearly reducing cap will force companies to use more and more of their stored credits to offset their emissions while the number of freely allocated credits will also be dropping. Eventually, companies will have to use their existing surpluses. The crucial word here is *eventually*, and the UK cannot afford to waste time waiting for this to happen, when more efficient options are available. As it stands, the surplus means that few companies will have to act on their emissions until 2016-18, only two years before Phase III ends (see Appendix 5 for the statistical evidence behind this, p.131). Hence while ETS supporters argue that the over-allocation has been resolved and will end once Phase III kicks in, the unappetising legacy will continue five years after this, an unacceptable delay. Their highlighting of this as evidence of the ETS's redeemable nature serves only to emphasis the fact that it is unworkable.

Two solutions have often been proposed to alleviate the credit surpluses actively, based on the dual factors of the recession and the NAPs: the adjustment of caps to reflect historical emissions for the former, and direct intervention for the latter. Neither is a sensible approach. The issue with adjusting the cap according to historical GHGs is that it will shackle industry's emission output to too low a level. By basing the cap on the period including the recession, where production fell dramatically, the average emission level will not reflect normal business conditions, as can be seen in Figure 1:1 (see p.XX). This will create a constraining ceiling on production at the highly vulnerable time post-recession when demand is picking up and output also needs to rise accordingly. Not allowing this growth potential will weaken EU national economies and cause extra-EU imports to rise. Indeed, enforcing a lower cap now will have a negative effect on their ability to reduce their future emissions: companies need to be able to regain their momentum and acquire the cash reserves needed eventually to invest in low-carbon technology during Phase III. Just because the cap was overly high before, does not mean that it should be overly low now. Two wrongs don't make a right.

The cancelling of unused allowances to reduce the glut and raise the price is an undesirable approach. This is perhaps more favoured by environmental groups, but has already been raised in the European Parliament once, and discarded. The 'Greenhouse gas emissions reductions and risk of carbon leakage' legislation was subjected to a vote on 5th July 2010, when the motion was rejected, with a majority of 87.[11] In it, policymakers had proposed 'setting aside 1.4 billion allowances from the EU ETS prior to 2020 as a possible solution to maintaining the incentives in the ETS and to guarantee the level of stringency foreseen at the time of the legislative procedure'.[12] This is a welcome outcome: proposals to set aside credits overlook the effect this has on investor confidence. The artificial price inflation would be contrary to companies' expectation and the EC's promise when the ETS began, that the free market decides the value of carbon. Moreover, once the precedent of intervention has been set, the potential for further intervention will overshadow certainty and further reduce confidence. If the scheme's reputation is tarnished, this would be reflected in the price of carbon. If trying to repair the ETS fails to steer it towards success, but towards another, unsatisfactory path, then it is better to be rid of it and replace it with something more effective.

Price fluctuations and short-termism

The large fluctuations in the price of EU ETS credits have undermined another cornerstone of the scheme: the confidence it was supposed to give low-carbon investors. In Phase I, the price per EUA varied between €30 and €0.03, and the price collapse was a consequence of the oversupply of credits. In Phase II, there is the possibility that the price of EUAs could see a similar fall *à la* Phase I if the oversupply continues or gets worse. As can be seen in Figure 0:1 (see p.xxv), the credit cost has been falling of late, and many pundits doubt this situation will improve. Barclays Capital conducted a reassessment of costs in July 2011 and found that the oversupply and reduced utility hedging meant:

> [Barcap] downgraded their average 2011 forecast for EU Allowances to 15.25 euros from 17.25; 2012 forecast to 17 euros from 24 and 2013 forecast to 23 euros from 30, based on the last price poll conducted by Reuters. The average price forecast for the period 2013 to 2020 was lowered to 30 euros from 40.[13]

Other pundits have cited different factors, including the threat of another recession, conflicting energy efficiency targets, the failure to move to more ambitious climate change targets and the diplomatic squabble over the EU ETS between Europe, China and the US amongst others. Whatever factors are blamed, it is clear that the credit price is subject to many conditions and is not necessarily predictable. For a market-driven commodity, this is understandable if undesirable for policymakers. This means that for Phase III, the only way might not be up, further reducing confidence.

Recently, the Parliamentary Environmental Audit Committee (EAC) has warned: 'the Government cannot place too much reliance on the price of carbon to drive investment in low-carbon technologies as the current price is too low and too volatile'.[14] The more fluid the price, the more risk there is in spending considerable sums on low-carbon technology, and few firms would consider investing in low-carbon technology with all the associated long-term costs if they suspected that carbon credit prices would soon fall. For the most part, reducing emissions is not a cheap process and one chemical firm has had to spend the huge sum of £400 per tonne of CO_2 reduced.[15] This is a significant cost and burden to the company, which is only manageable if it can sell surplus permits at a reasonable price and sees non-investing rivals also straddled with the costs of obtaining permits. Were the price to collapse, this would put the low-carbon firm at a considerable competitive disadvantage compared to rivals who merely bought credits the entire time. The ETS has to navigate a path between the Scylla and Charybdis of too high or low a price, but without a rudder, a consequence of its market-based nature. This uncertainty plagues the ETS, and many companies feel they are wasting time and money through the need to mitigate the issue. A report by the Carbon Trust has found that 40 per cent of FTSE 100 companies do not have a clear plan for how they will reduce their emissions in the future because they either never had any targets or their existing ones have expired.[16]

Those companies without plans are technically in the minority, and supporters of the ETS have looked to evidence such as the statistic that 32 per cent of companies have found 'the EU ETS has increased the importance of CO_2 emissions and energy efficiency at the board-level', to prove the System is working. However, it is important to note that this is only true in the short term.[17] Of those with plans in the Carbon Trust survey, 55 per cent only last until 2013, 18 per cent until 2020, and just four per cent plan beyond this.[18] This is primarily due to the confusion about what will happen in Phase III, as while companies know they will receive fewer allocated free credits, the size of this reduction is not always obvious. A similar survey by Point Carbon found that 37 per cent of companies are either 'very uncertain' or have 'no idea' how many free EUA's they will receive in Phase III and so are unable to plan for its implementation. Just eight per cent said they 'knew exactly' what their allocation would be.[19] Only 13 FTSE companies have taken Phase III into account. Given the pace of change in the EU ETS, this is unsurprising: it is very difficult to formulate very far ahead, so firms are unwilling to set out long-term strategies. Even those companies that are acting on the EU ETS are not doing so in the way the policymakers were hoping for: the low cost of permits fails accurately to incentivise or reflect the cost of investing in major low-carbon enterprises. Firms that decide to invest in real low-carbon technology when they could implement other, easier short-term cuts will effectively be penalised for doing so.

The volatility is in part derived from the short-term nature of the ETS. Its running has only really been planned carefully until 2020: after that date, the scheme will continue, but in what form, or whether this will still be Phase III or Phase IV is entirely unknown. Given this date is less than nine years away, the majority of affected companies have business plans reaching to future dates where this uncertainty will become a serious issue and, at the moment, there is no way to tell whether a large-scale investment will pay off. Without this knowledge, many will find buying credits the cheaper option. To be fair, this short-term bias is not just the fault of the ETS, and all climate change policies are reliant on successive governments maintaining a constant approach and equally valuing it as an issue. This can never be guaranteed.

In terms of the ETS, it will be hard to re-engineer the scheme to take longer-term trends into account. While it is obviously designed to decrease emissions, it actually requires a baseline of constant emissions to keep the credit price steady. If every company took their obligations as seriously as possible, and did find ways to reduce their emissions, then there would be no demand for credits on the carbon market, which would bring back down the price of credits. Of course, if everyone were that angelic, no one would need to purchase credits by this point anyway. However, not every company will be this responsible, but assuming that over time, around 2020, the price of credits does cause many companies to act, the price would then fall as they do and those that avoided making the investment and continued to just buy credits at auction would no longer be penalised. The tightening of the cap is supposed to avoid this, but is a cumbersome technique. Overall, the successful reduction of emissions can therefore actually reduce the long-term incentives to invest in low-carbon technology. Similarly, those selling their saved credits would receive less for them and the lower the price, the lower the incentive to reduce their emissions beyond the cap. The EU ETS cannot take this long-term shift to low-carbon technology into account fully and, by its very nature, undermines its own aims.

Banking allowances, windfall profits and the power sector

While much has been said about caps being tightened in Phase III, to avoid the over-allocation problem so far experienced, this will be undermined by the ability to carry allowances over from Phase II to Phase III. The majority of allowances have thus far been given out free to industrial companies and, by the end of Phase II, it is estimated that €18 billion's worth of EUAs will have been handed out to them. The money being given to the companies in the form of these EUAs has to come from somewhere, and this is primarily the domestic consumer, with each EU citizen effectively paying €36 for the windfalls.[20] Not all these credits have been used and, in 2009 alone, 70 per cent of installations benefited from receiving more allowances than they required.[21] This means that there has been little reason for these sectors, whether at risk of carbon leakage or not, to invest in low-carbon technology. The problem here is that the credits have no expiry date, and this means spare credits can therefore be 'banked' by installations to be used at a future time, i.e. to meet the targets when they get tougher post-2013. An estimated 123 million credits will be transferred.[22] This gives companies a head start and leeway to adjust, but provides no environmental benefits. In addition, the inception of auctioning will be slow. It will take seven years for the 30 per cent auctioning in 2013 to reach 80 per cent. In the meantime, industries will continue to receive free allocations, albeit at a lower level, and some could find themselves having to make no emission reductions until after 2020.

The lack of an expiry-date means the credits guarantee the CO_2 emissions *will* happen at some point. This undermines the prime advantage cap-and-trade schemes are supposed to supply, that the level of present emissions is always known. Instead, all the EU now knows that it will have a level of future emissions far greater than it should be. During Phase III, this might mean that the level of emissions fails to fall in line with the cap, at least until the banked credits are all gone.

There are two ways in which companies make windfall profits of their freely allocated credits. They can either pass the 'cost' of the credits on to consumers, in the form of higher prices for their products, or sell the free credits on the carbon market if they are not required. Some companies find themselves in the position to do both. This is not a problem likely to go away in the near future, even once Phase III occurs, thanks to the ability to bank surplus credits. This means that, overall, some companies are making significant windfall profits for doing nothing or very little about their emissions. At best, this is inefficient and at worst, this is subsidising companies' continued use of carbon-intensive technology, the antithesis to the scheme's intention. In 2009, a quarter of the surplus credits were in the hands of just ten companies and, as of 2011, they have received credits to the value of €4.1 billion during Phase II, four times the EU environmental budget of the whole period. Already, ArcelorMittal and Lafarge have made significant earnings from their allocations, selling €172m and €300m of EUAs respectively.[23]

While it has been hoped that Phase III would put an end to these windfall profits via increased auctioning and benchmarking, this does not appear likely in the Phase's first years. CE Delft conducted an investigation into this:

> The situation in Phase 3, however, will change, as a larger share of emissions for energy-intensive industries will be auctioned. The impact of auctioning for firms that fall under the scheme of transitional free allocation will be limited as these are only responsible for a very small amount of total emissions. However, the impact of benchmarks is much larger. There is to our knowledge currently no study that has estimated the total amount of emissions that will fall under a benchmark. Visual inspection of, for example, the cement study by Ecofys (2009) shows that less than 10 per cent of total emissions in the cement sector will, finally, be auctioned. So one may conclude that only a small fraction of total emissions for industry will, after 2013, fall under an auctioning regime.
>
> As we would expect that companies still would pass through the opportunity costs of their freely obtained allowances in Phase 3, total windfall profits will be diminished by the small amount of allowances that will fall under an auctioning regime.

Overall though: 'This study has shown that windfall profits are likely to occur as long as emissions are allocated free of charge.'[24]

Moreover, as it is assumed that the price of allowances will increase over time, companies have an incentive to retain allocated allowances now in order to sell them later for a greater profit. This has led to many companies holding on to their allocated credits and buying Kyoto credits instead, whose price, while slightly lower, is unlikely to rise as much in the future. The UN-backed credits are then surrendered instead and there is no need to dip into stores of EUAs until the price is attractive enough to sell them all. The difference between these two sums is then a pure profit for the company, with no rise in low-carbon investment. It has been estimated that 28 per cent of ETS installations have taken advantage of this loophole, reaping €628 million from the difference.[25] This practice, while morally dubious, is entirely above board and legal within the framework of the EU ETS, despite the fact there is no benefit to anyone except the company itself.

The problem is worst in the power sector. The industry was given stricter targets to meet, to balance out the more lenient targets for energy-intensive industries, on the assumption that there will be few competitive issues. In other words, if EU production costs rise, while it is possible to import steel from China, it is not possible to import electricity. It was therefore seen as entirely reasonable to allow the power firms to be exposed to any increase in the cost of ETS credits and to pay for all their necessary credits come Phase III. This would make sense if the companies were paying for their credits from their own pocket, but they are not. It is worth mentioning that while the power sector will see full auctioning from 2013, this might not eliminate all windfall profits: a significant proportion of the allowances will be made up of CERs (see p.18), which are, on the whole, cheaper than EUAs, so companies will pass the cost of the more expensive EUAs on and pocket the difference.

At present, these companies are passing on the 'cost' of their free allocations to consumers by raising energy bills, thereby earning their own form of windfall profit. By the end of Phase II the eventual gains have been estimated at between €16 billion to €50 billion.[26] This has been a real boon to the fossil fuel-based energy market and has sustained status-quo investor confidence: it has shown there is still less risk in pursuing 'business as usual' than venturing into low-carbon power sources, as added marginal costs simply fail to erode profit margins. The use of past emission levels to decide the number of credits given to each power installation has meant that, even within the fossil fuel market, those most 'rewarded' were the most carbon intensive power companies, who emitted the most and therefore generated the most permits. In total, it actually paid for power companies to maintain the worst carbon emissions possible.

... these companies are passing on the 'cost' of their free allocations to consumers by raising energy bills, thereby earning their own form of windfall profit

This surprising situation is warped further given the ETS only covers carbon-intensive power generators, leading to the perverse situation of carbon-neutral power sources, such as nuclear and renewables, receiving no credits, despite their obvious environmental benefit. While renewables receive subsidies from elsewhere, and therefore do not need to be included in the ETS, it still stands to reason that a company now switching from fossil fuels to low-carbon power would be rewarded with credits to sell, while those that made the shift as early as possible, and therefore bore the brunt of the first-mover risk, have been ignored. Once 100 per cent power sector auctioning also begins in 2013, the ethos will shift from rewarding to punishing, as installations will have to buy all their allowance. Given the continuing ability to pass this penalty cost to consumers, and the lack of credits to sell on, enthusiasm for investing in carbon-reduction technology will further decline. The EU ETS fails to ensure that low-carbon investments become less risky and therefore does little to make them appear more attractive.

The UK will subsidise improvements elsewhere to avoid making its own

The EU ETS aims to improve the GHG emissions of all European countries. There are no specific benefits or improvements to the UK which will enable it to meet domestic targets. It is for this reason that other unilateral measures and costs exist, such as the carbon price floor and Climate Change Levy. Given companies can easily buy their way out of reducing emissions, and wealthier companies are more readily able to do this, many British ones will simply choose this easy route. Ultimately the scheme is not about paying to emit, but about reducing CO_2 output, and anything less than this can be deemed a failure. The UK Government has already estimated Britain will be a net purchaser of 14-25 million permits, therefore ignoring the real *raison d'être* of emission reductions.[27] For the UK at least, the ETS will be a failure.

If the UK is buying permits, this means they are being sold by other countries with a surplus of credits. Of the 30 participating countries, many are worse emitters with easy emission reductions still possible. This means that they are likely to reduce their emissions with low investment costs. As such, Britain will be subsidising them to become as efficient as the UK already is, while emitting more itself. Alternatively, if Britain does start to reduce its emissions dramatically and creates a surplus of credits, the result is not necessarily beneficial to the environment: it merely allows another country to emit more. The balancing act of the buying and selling of credits fails to reconcile the difference between individual installations reducing their emissions and overall emission reductions. In theory, emissions should be reduced as a whole across Europe, but again this oversimplifies the issue. The introduction of the Clean Development Mechanism (CDM) drags extra-EU countries into the mix, and, just like Britain on the smaller scale, the whole EU pays non-member countries to reduce their emissions so that it doesn't have to.

The Environmental Audit Committee has noted that: 'even taking reduced economic output due to the recession into account, the largest cause of the reduction [in emissions since 2007] is the EU ETS itself encouraging greater use of gas in power generation'.[28] In the European context, easy cuts are not the desired outcome either, as they do not involve any low-carbon investment. The cap is sufficiently high and the credit costs sufficiently low that they will be satisfied by these non-dramatic changes, giving some sectors a much easier ride than others.

CHAPTER 2

The Failure of the Clean Development Mechanism

- The Clean Develop Mechanism's (CDM) credits, CERs, are worth the same as EU ETS credits and can be submitted by ETS installations instead of EUAs. CERs are generated by extra-EU emission reducing projects to be sold on, to incentivise green investment, especially in developing nations. The EU is effectively offloading its ETS obligations in a 'do as I say, not as I do' move.
- The CDM is a 'zero sum' mechanism. For example, a CDM project reducing emissions by 1,000 tCO_2e will generate 1,000 CERs, which can be bought by ETS installations to allow the emission of 1,000 tCO_2.
- The CDM is vulnerable to corruption. A study of the top five UN-accredited CDM validatory bodies found that on a scale from 'A' (very good) to 'F' (very poor), none scored higher than 'D'.
- A 4,000MW coal plant in Gujarat, India, has received CERs because it is marginally less polluting than other coal stations. This is despite the fact it emits 26 million tonnes of CO_2 per annum, will do so for at least 25 years, is India's third largest source of emissions and is the 16th largest worldwide.
- Industrial gas credits reap huge profits. HFC-23 generates 11,700 credits per tonne destroyed at approximately €12, but costs only €0.17/tCO_2e to destroy: a 7,000 per cent markup. As a result, some companies are creating HFC-23 just to destroy it in order to generate credits. If the scheme did not exist, these emissions would never have been produced.
- This is especially rife in China where, because it is so lucrative, the government taxes CDM revenues at 65 per cent, expecting to generate £1.7 billion by 2013.
- While gas credits have been banned from May 2013, lobbying led to a delay in the ban and 412 million credits are still waiting to be issued through the scheme.

At the same time as the EU ETS was being set up, the Kyoto Protocol also came into effect, in February 2005. Despite being designed to run independently, it was decided to incorporate Kyoto based credits into the EU ETS, to help other countries reduce their emissions as well. This became known as the 'Linking Directive' and allows EU member states to use these credits to cover a percentage of their emissions. However, this practice is at best flawed, and at worst exploitative and actually damaging to the environment.

There are two main forms in which Kyoto credits come: the Clean Development Mechanism (CDM) credits, known as the Certified Emission Reductions (CERS), which form the bulk of the non-EU credits used in the ETS, and Joint Implementation (JI) credits, known as Emission Reduction Units (ERUs). The former is designed for projects in the developing world, and the latter for those in other industrialised countries. The Kyoto mechanisms allows emission reducing projects to generate credits equivalent to the amount of CO_2e reduced, and these can then be bought by industries in developed nations. Both of these are the equivalent of one tonne of CO_2: in other words, they are worth the same amount as an EU ETS EUA credit.

In total, the EU ETS allows half the emissions reductions of the 2020 target to be met through these CERs, which can be submitted by companies in lieu of EUAs and this will continue into Phase III, thereby reducing the pressure on a company to invest in low-carbon technology.[1] This burden is further lifted by the fact that CERs are significantly cheaper to purchase than EUAs, as can be seen in Figure 0:1 (see p.xxv) meaning many companies prefer to buy these, and save their allocated EU credits for later needs. Indeed, there are fears that because of the expense saved, industries will be more likely to continue to use carbon-intensive technology, making any eventual switch all the more expensive and unlikely. This can be seen in Spain, which will have doubled its domestic CO_2 emissions from 1990 through to 2012.[2] Throughout Europe, the use of CERs has been steadily rising, and 2010 saw 117 million of these certificates used, a rise of 50 per cent on 2009 levels.[3] In addition, 20 million ERUs were used, making a total of 300 million Kyoto credits used to date, with a value of €3.9 billion. Figure 2:1 (see p.19) shows just how prevalent their use is.

At times though, the CDM effectively legitimises carbon leakage and is at best a 'zero sum' game with no global benefit. While local installations might reduce their emissions, this allows the buyer of CDM credits to raise theirs. For example, if one firm in a CDM country halves its emissions from 1,000 tCO_2e to 500 tCO_2e, then 500 CERs are generated, allowing another company in an industrialised country to pollute the otherwise saved 500 tCO_2e, delivering a smaller benefit than it would appear to at face value. In some

Figure 2:1

Total ERUs and CERs surrendered by selected EU ETS participants until 2010

Source: European Environment Agency, EU ETS data viewer

cases, the outcome is entirely negative, as in the case of industrial gases discussed below (see p.22). Here, the emissions are deliberately produced and then destroyed to generate credits, which then allows EU firms to emit more CO_2. In other words, without the CDM, these emissions would never be made. The EU never intended for the CDM's inclusion to deliver further emissions reductions beyond ETS requirements, but to provide companies with a cheaper way to meet their obligations. However, it is clear that this aim is not necessarily being met anyway.

While adding to the baseline of EU emissions, CDM projects might not actually reduce emissions in a developing country either, but merely slow the speed at which they are increasing there. CERs are only supposed to be provided if the project could not otherwise go ahead, but in many cases, CERs are being provided regardless of this, resulting in a net increase in emissions. Effectively from the EU perspective, the mechanism revolves around the principle of 'do as I say, not as I do': it allows prosperous EU states to export their obligations to developing countries, rather than acting on their own responsibilities, and they pay less than they otherwise would. This is hardly proof that the EU is leading the world on curbing carbon emissions and suggests that these richer nations are using their financial clout to get others to do their work for them – a form of neo-imperialism.

That the EU has imposed an upper limit on half of mandated reductions being able to come from outside Europe is not impressive. The UK is as guilty of this as any other country, having submitted 17 million CERs and ERUs from 2008-10.[4] Worryingly, the UK government is pressing for more widespread use of these, and wants to keep 55 million to use to meet future non-ETS targets from 2013-17 such as the UK specific carbon budgets. At a meeting of the Environmental Audit Select Committee, Caroline Lucas MP pointed out that retaining these is a form of 'get out of jail free card', as the offsets allow the UK to buy its way out of failing to meet goals domestically. In response, Chris Huhne, Secretary of State for Energy and Climate Change, said: 'If we can meet targets domestically, we should. But [economic] forecasts are enormously uncertain. We think it makes sense to keep that flexibility [in using offsets].'[5]

The CDM provision does not actually specify *who* is making the reduction, whether it is the actual reducer or the purchaser of the CER. In all likelihood, both will claim the acknowledgment and put the same reductions towards their own targets, so global emission reductions might not be as steep as national statistics might suggest. Baoshan Iron and Steel Co., a Chinese government-owned company, claimed it reduced emissions by 63 per cent and 1.61Mt CO_2e of this generates CERs.[6] However, another company in

Europe would also be counting the buying of these CERs and the same reductions as progress towards their target. This is bad enough, but it also gives countries such as China the excuse, with some justification, to refuse to agree to any further international emission targets. Given they are already investing in low-carbon technology on behalf of other European nations, why should they need to do anything more? This has an economic impact on the EU, as China will then continue to maintain its low-cost competitive advantage, at a time when further green costs are being piled on European businesses.

Corruption within the Clean Development Mechanism

The CDM market has been rocked by various scandals, an unsurprising by-product given how lucrative it is and how little independent regulation there is. Projects have been poorly managed, failing to deliver the promised reductions or overemphasising the level of reductions in order to generate the maximum number of CERs. This has significant consequences for the EU ETS because almost all of the CDM credits end up within the System, undermining the emission cap. Moreover, its corrupt practices drag down confidence in the ETS and consequently its ability to deliver emission reductions. Given the amount of money flowing from the ETS into corrupt practices, the CDM is something akin to an environmental black hole, swallowing money but failing to have any real benefit.

This shoddy behaviour is rife throughout all levels of the CDM process. Before receiving credits, any potential beneficiary has to be vetted and audited to ensure they are eligible and to decide how many they should receive. However, verifiers and validators are not performing as well as they should be. A study of the top five UN-accredited validatory bodies found that on a scale from 'A' (very good) to 'F' (very poor), none scored higher than 'D'.[7] In addition, two firms that validated between them nearly two-thirds of all offset schemes have both been suspended for irregularities in staff and accounting.[8] The problem is that consultants earn large fees for securing CDM status, and rather than delivering legitimacy, many now rubber-stamp projects while some even fabricate claims to deliver CERs. A typical ploy is to portray a scheme as economically unviable on its

A study of the top five UN-accredited validatory bodies found that on a scale from 'A' (very good) to 'F' (very poor), none scored higher than 'D'. In addition, two firms that validated between them nearly two-thirds of all offset schemes have both been suspended for irregularities in staff and accounting.

own but profitable if CERs are generated, ignoring the myriad of other tax breaks and subsidies that might be offered by the corresponding government. Indeed, with three-quarters of applicants applying for CERs while nearing completion, not before the project is started, it is clear that the Mechanism is just seen by some as a cash-cow.[9]

Non-renewable or carbon-intensive power sources can also apply for CDM status. A 4,000MW coal plant in Gujarat, India, has received CERs because it is marginally less polluting than other coal stations. This is despite the fact it emits 26 million tonnes of CO_2 per annum, will do so for at least 25 years, is India's third largest source of emissions and is the 16th largest worldwide.[10] Subsidising high-carbon power, simply because it is more efficient, will directly impede the development of low-carbon investments: it will strengthen the current high-carbon technological lock-in.

The worst offenders cynically manipulating the CDM are those involved with generating industrial gas credits, from HFC-23, a refrigerant gas and N_2O, a by-product of nylon manufacturing. These are two of the most potent GHGs: respectively N_2O and HFC-23 are 310 and 11,700 times more damaging to the atmosphere than CO_2 and therefore earn 310 and 11,700 credits per tonne of N_2O and HFC-23 destroyed. HFC-23 costs only €0.17/tCO_2e to destroy, but companies are receiving approximately €12 at the current market rate: a 7,000 per cent markup![11] As a by-product from creating refrigerant gases, companies that destroy their output of HFC-23 have found they can earn over double from CERs compared to what they would receive from selling the refrigerant gases commercially.

...some firms, especially in China, now create refrigerant gases just to create the HRC-23 that they then destroy, to reap the rewards. Indeed, of the largest ten generators of CERs, seven are HFC destruction facilities, showing just how lucrative a system it is

Little wonder then that some firms, especially in China, now create refrigerant gases just to create the HRC-23 that they then destroy, to reap the rewards. Indeed, of the largest ten generators of CERs, seven are HFC destruction facilities, showing just how lucrative a system it is. Having cottoned on to this, the Chinese government has also been making a significant sum out of the scheme. It charges a 65 per cent tax rate on HFC-23 CDM profits, which goes into a supposed 'sustainable development fund', the purpose of which is unclear. It has been estimated that, by 2012, the Chinese government will have generated $1.7 billion from the tax.[12]

Figure 2:2
Number (%) of CDM projects in each category

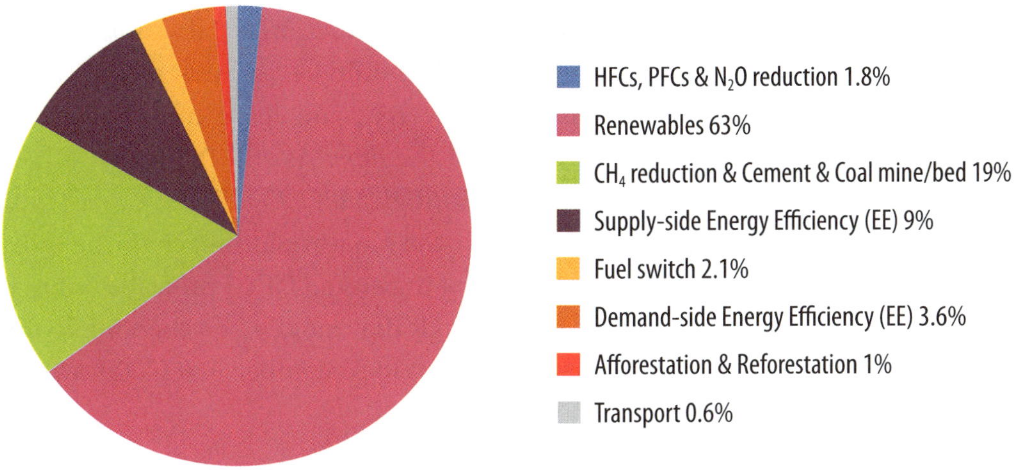

Source: UNEP Risø Centre, CDM Pipeline database, Analysis, Chart 3

Figure 2:3
CERs issued in each category

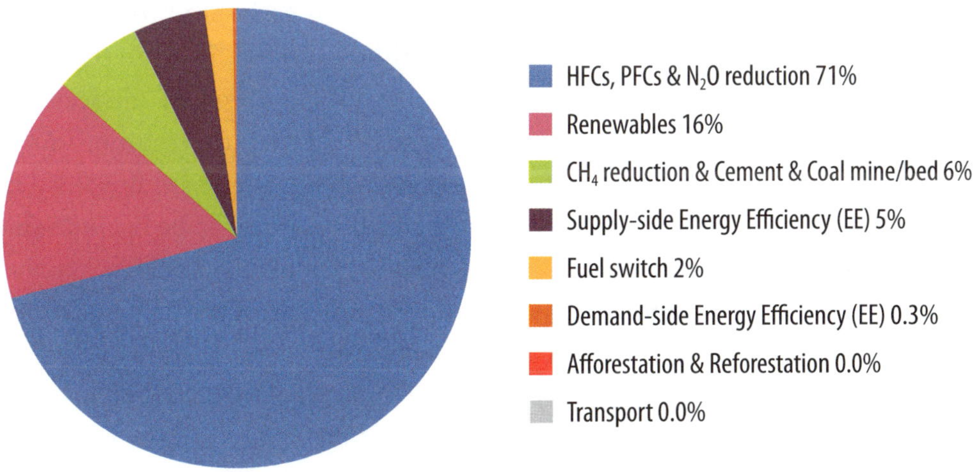

Source: UNEP Risø Centre, CDM Pipeline database, Analysis, Chart 6

The industrial gas credit racket is not just beneficial to the companies making a profit. It also benefits indolent European countries who gain a constant source of new CERs to buy, avoiding the need for making low-carbon investments themselves *ad infinitum*. Overall, 77 per cent of EU ETS CERs come from the reduction of HFC and N_2O and, in Phase II, these credits amounted to €2.8 billion.[13] It should be said that not all companies are directly investing in these gas eliminators directly, and some come into gas credit ownership via the carbon credit exchanges, but the effect is still the same. It has been estimated that by 2012, total CERs generated through HFC-23 will have been worth €6 billion.[14] Figures 2:2 and 2:3 (p.23) clearly show the bloating influence of the industrial gas producers: they are a tiny fraction of the total registered CDM projects, but make up the vast majority of credits issued. It is bad enough that money has been made through ransoming the environment, but it is also important to remember that CDM scams draw money away from projects that genuinely need funds.

This absurd situation developed despite Connie Hedegaard, the EU's Climate Action Commissioner, describing these industrial gas credits as having a 'total lack of environmental integrity' and opposing their use.[15] This eventually meant the subsidy of such facilities was banned by the European Commission in January 2011, but this will only take effect from May 2013, as a result of intense lobbying pressure from various organisations. Originally, the date had been set for January 2013, and while this window of a few months appears small, it has been estimated that 52 $MtCO_2e$ of industrial gas credits could be used during that time in the ETS, more than the entire annual reduction of 35 million tonnes required in 2013.[16] Overall, the long window remaining for their continued abuse during Phase II has the unwelcome side-effect of accelerating their use, as companies attempt to take advantage of the credits while still possible: there are still an estimated 412 million credits waiting to be issued which will no doubt flood the CER market and pull down the price nearer to their expiry date.[17] As a result, the Chinese government has been approving projects at a record rate. From mid-July to mid-August 2011, 86 projects were approved, beating the average of 51 per month previously and 2009's average of 47.[18] This ethical quagmire has led to the launch of 'green CERs' by the ICE carbon exchange. Maturing in December 2013, these carry roughly a 90 eurocent premium over the 2012 CER, and they do not include industrial gas credits. Their name alone suggests how ludicrous and widespread CDM manipulation has become. Moreover, there are also fears that the credits could re-enter the carbon market again under bilateral deals between developing/developed countries, or through a black market CER trade: according to the Environmental

Investigation Agency, the second largest CDM HFC-23 project, Shandong Dongyue Chemical Company Ltd, 'has previously been implicated in the illegal trade in ozone-depleting substances'.[19]

It would be wrong to tar all CDM projects with the same brush, but given financial aid is being provided to schemes ill-fitting with CDM ideals, there must be a proper reassessment of the Mechanism's scope. Companies in China and India especially are acting the cuckoo and crowding out other, genuine low-carbon projects that require funding to take place. Figure 2:4 (see p.26) shows the domination of the four countries responsible for the largest generation of CERs. Between them, they now account for 72 per cent of all CDM projects. This suggests that the least developed countries, those most in need of CDM credits, are not being provided for. Indeed, the entirety of Africa has generated only 3.6 per cent of all CERs issued.[20] Even some worthy projects are erroneously receiving credits, as the improvements would have been made anyway so there is no 'additionality'. While there are too many CERs flying around and making cynical manipulators a substantial profit, in a way, there are still not enough to go around as not every deserving initiative can obtain them. Clearly, the solution is better scrutiny of CDM projects. The situation is all the more disappointing given that the legislation behind the EU ETS has a clause stating that CERs can be allocated on a quality basis: 'high quality CDM credits from third countries should only be accepted in the Community scheme'.[21] What it is really lacking is the ability to revoke credits handed out to dubious projects: even if the UN realises a venture has falsified emissions reductions, nothing can be done once the CERs are handed out, despite the fact it will mean another firm is using the ill-gotten CERs to pollute more themselves and increase net global emissions.

Companies in China and India especially are acting the cuckoo and crowding out other, genuine low-carbon projects that require funding to take place

It has been estimated that there are enough CERs (1.6 billion by the end of 2012) for installations to meet their targets for Phase II without actually reducing their own emissions at all.[22] Similarly, the Phase III 21 per cent ambition is unlikely to be met within Europe. The Environmental Audit Committee called for a tougher cap to reflect this, but punishing industry for the inherent faults of the System is unfair.[23] A more sensible approach would be to restrict the number of CER offsets available, to re-focus companies on reducing their own emissions rather than that of others. The CDM scheme is expanding globally and has spiralled out of control. Validators

Figure 2:4

All CDM Projects in the pipeline in the four biggest CER producing countries, as a fraction of all projects

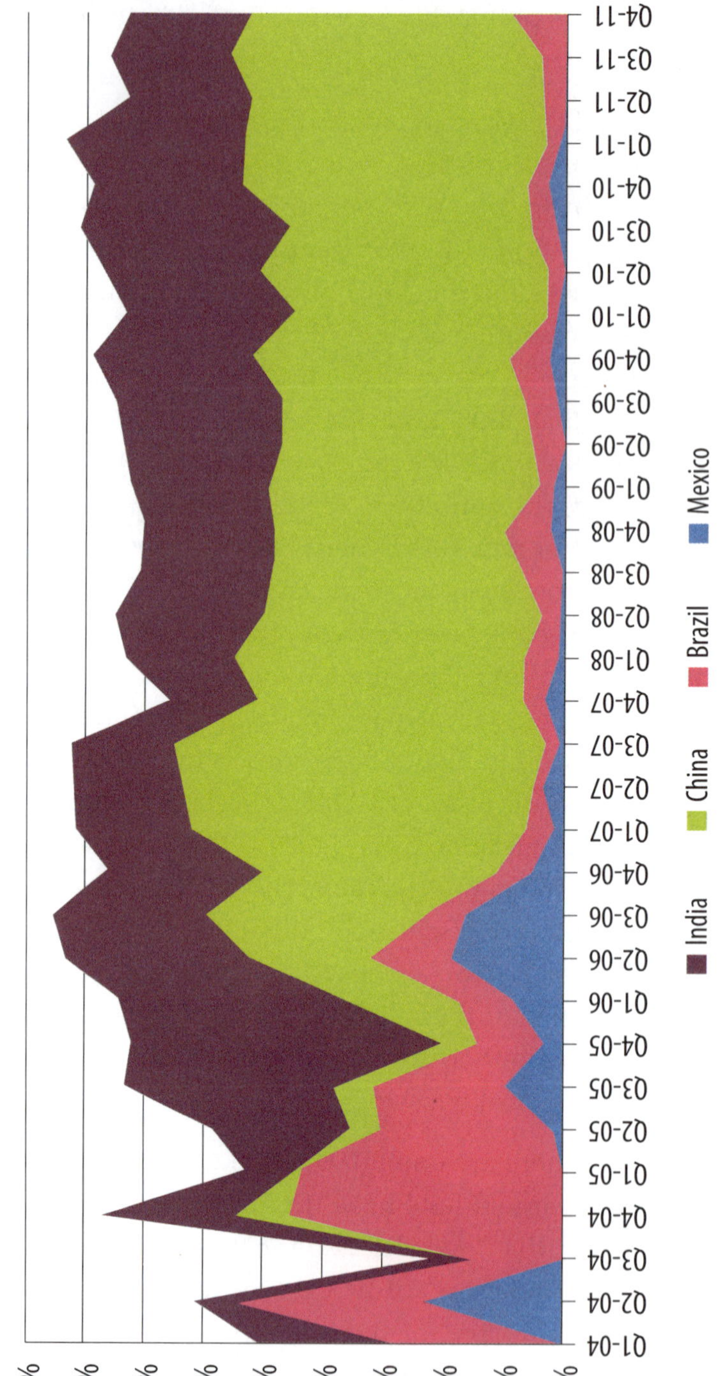

Source: UNEP Risø Centre, CDM Pipeline database, Regions, Chart 11

are being paid for delivering results, not impartiality, and this has led to a 'wild west' of dodgy projects being approved. The whole system needs far more transparency and policing: establishing an independent moderating body such as the UN would be a solid start.

CHAPTER 3

Profiteering from Good Intentions

- Power companies pass the cost of the EU ETS, real or not, onto consumers via bills. In 2009, the average UK domestic electricity rate-payer faced paying an extra £24 because of the ETS, equivalent to almost a third of all imposed environmental costs. This will increase substantially in Phase III as full auctioning for the power sector is implemented.
- A UK poll conducted by Populus found only one per cent of consumers would willingly pay £500 in green costs on energy bills. Ofgem now estimates that total green costs to energy consumers by 2020 will be between £250 and £600.
- Carbon trading is the fastest growing commodities market and demand for ETS credits is driven by commercial banks, not installations: a trial of non-competitive auctioning was discontinued due to lack of interest.
- Financial institutions have developed a vested interest in the EU ETS, with a massive conflict of interest. Banks own or invest in offsetting firms and the higher the price of carbon allowances, the more demand there will be for carbon offset credits. This incentives them to use their size to buy ETS credits to push up their price.
- Banks successfully bully the ETS to maximise their profits, for example by delaying the industrial gas credit ban. Even the World Bank lobbied for this. It has investments in two major HFC-23 schemes worth around €1.6 billion.
- Crime in the EU ETS is rife, and 90 per cent of all market activity in 2009 was estimated to be criminal by the European Law Agency. Carousel fraud alone is estimated to have cost the EU €5 billion in lost tax revenue.
- Plans to increase the security of the system will amount to nothing if adherence to the heightened security is voluntary.
- The UK Government expects to receive £50 billion in ETS revenues by 2050, but has refused to hypothecate this to environmental projects. The ETS thus and gains the appearance of a stealth tax.

The cost to consumers

It is a fact yet to be accepted by most green groups that the existing methods of investing in decarbonising the economy, via green taxes on the producers and consumers and subsidies paid for by the taxpayer, *will* lower the standard of living for much of the population. Depending on the personal opinions and ideologies of the reader, this may or may not be deemed an acceptable situation, but for mainstream politicians, reducing the quality of life of the electorate is not normally a vote-winner. To be acceptable as a whole, it needs to be demonstrably beneficial to future generations, in terms of energy price, security or side effects.

This is as much true for the EU ETS as for any other green policy. Domestic users and households are already being increasingly subjected to hefty energy bills, inflated by environmental taxes and levies, and the ETS forms an increasing part of this, despite the fact that households are not industrial emitters of CO_2. From the viewpoint of the homeowner, Britain has seen the worst energy bill increases throughout Europe: from 2004-09, domestic electricity prices rose by 75 per cent, and gas prices by 122 per cent.[1] In total, according to DECC, environmental charges currently make up £70-90 of average bills, of which the EU ETS is just one.[2] In terms of the EU ETS, the charges that bill payers face are the derived costs imposed by energy companies on bills to compensate for the higher costs the companies themselves supposedly face in generating the electricity. As the cost of carbon credits rises, so too will the burden of the EU ETS on domestic bills. In 2009, the average customer faced paying an extra £24 because of the system, equivalent to almost a third of all imposed environmental costs.[3]

From the viewpoint of the homeowner, Britain has seen the worst energy bill increases throughout Europe

It is important to note that Phase III will have no positive effect for the consumer. As research from the Pew Center on Global Climate Change shows, auctioning will be no better or worse:

> Since every allowance used to cover emissions means the loss of the opportunity to sell that allowance, an opportunity cost is incurred and that cost is the foregone market price for allowances. Accordingly, whether allowances are distributed for free or through an auction will typically have no effect on market prices in competitive electricity markets, although it will affect individual supplier profitability.[4]

All that matters is therefore the price of the carbon credit, which will almost certainly rise steadily throughout Phase III: Standards & Poor's has estimated that wholesale energy prices could rise by up to a further 20 per cent due to this.[5] Whether the power companies are enjoying windfall profits or just breaking even, in both phases the consumer loses and is used to recoup the marginal cost of the ETS.

It seems to have escaped the notice of the EU ETS architects that consumers are less concerned about being green than they say they are. In 2008, while 62 per cent of EU respondents felt that global warming and climate change is the 'most serious problem currently facing the world', only 44 per cent were actually willing to pay premiums for green energy, with only two per cent willing to accept an increased cost of 30 per cent or more, something highly likely closer to 2020.[6] Given this survey occurred prior to the recession, already squeezed consumers are unlikely to feel any more positive now about higher bills. The consumer angle is a serious one, not least because, as their EU ETS based costs continue to rise, public support for the scheme will decline. In time, this could reach the point at which public opinion values environmentalism below their ability to travel freely and keep warm. If this happens, and voters turn against the regulations, this could force politicians to unravel not just the EU ETS but the whole green agenda, undermining all efforts towards a low-carbon future. This tipping point is not that far ahead. A UK poll conducted by Populus for Centrica found that while a third of respondents were willing to pay £100 extra to lower carbon emissions (and therefore two-thirds would not), only one per cent would willingly pay £500. Ofgem now estimates that total green costs to consumers to meet the power sector's 2020 commitments will be anywhere between £250 and £600.[7] Would-be low-carbon investors are aware of this, and accordingly have to take the risk of policy reversal into account. It would be far better not to jeopardise the whole green agenda by rushing into renewable power and emission reduction, and instead to rethink how long-term goals can be met with the minimum cost to the domestic user.

It would be far better not to jeopardise the whole green agenda by rushing into renewable power and emission reduction, and instead to rethink how long-term goals can be met with the minimum cost to the domestic user

The financial sector's exploitation of the carbon market

The EU ETS carbon market was designed to facilitate easier compliance for obliged installations by acting as a forum in which credits could be bought and sold. However, this practical *raison d'être* has been overshadowed by financial institutions harnessing the market as a financial investment opportunity and worse, and this is practically endorsed by the EU. Carbon trading is the fastest growing commodities market and, like most trading markets, it is highly complicated. With its confusing regulations and opacity to outsiders, this means it is ripe for exploitation by financial service companies who can offer to buy or sell allowances and profit themselves. With hundreds of billions of pounds' worth of carbon trades already made, speculators are able to use pollution to make a quick buck with no care about reducing emissions.

The profiteering out of the EU ETS is a systemic problem, most frequently seen at the low-level auctioning that has already taken place in Phase II, and the UK has experienced this as much as any other country. The first British auction in November 2008 saw £54 million's worth of credits sold at a clearing price of €16.15. Demand was four times greater than supply due to the involvement of Barclays Capital, BNP Paribas, JP Morgan and Morgan Stanley, who were acting as 'primary participants', that is, the intermediaries through whom EUA users have to act in competitive auctions. In return, the primary participants, who undertake commissioned bids for no fee, are allowed to buy credits for their own purposes as well.[8] This was a fight between middlemen, not EUA users. In September 2009, a new non-competitive bidding facility was initiated, as a means for operators to buy their allowances in order to meet their targets. This facility was discontinued in January 2011 due to the low demand for non-competitive auctioning and only the competitive process remains active. This strongly suggests that the majority of EUA auctioning activity is not happening for the benefit of installations or the environment, but for the self-interest of the banks. For them, the value of being a primary participant is that it allows them to bid on their own behalf as well.[9] It is unlikely this has escaped the notice of the government, especially since VAT is charged on the credits at auction, implying they are more of a financial commodity than a means to save the world. The low demand for non-competitive EUAs is likely to be due to the continuing over-allocation of free permits, but this means existing auction stocks are being bought for their profit potential in years to come, when demand will be higher. In this way, financial institutions will command control over their price, given the credits roll over, and they will be holding much of the spare capacity, pushing up the price of credits in Phase III, to benefit their profit margins at a cost to the genuine ETS installations.

The domination of banks as intermediary 'primary participants' is not what the Department of Energy and Climate Change (DECC) hoped for. The advantage of acting as a middleman was that: 'Primary Participants can satisfy their demand for allowances directly, while potentially gaining reputational benefits, for example enhancing their green credentials.'[10] This was also noted in feedback from indirect permit buyers after the

2008 auction, who noted that the pool of the four primary participants was too small and the burdens required of becoming one themselves prevented them from buying allowances directly.[11] Joining just to profit out of the environmental project was not a desired outcome, and all the current nine primary participants are banks. The problem might not be a large one at present, with such a comparatively small volume of EUAs being auctioned, but during Phase III, when this will rapidly increase, the big banks stand to dominate the UK auctioning market, crowding out any newcomers via their superior knowledge and experience of the processes. Unless a cartel of avaricious middlemen is acceptable, the methodology of auctioning needs to made much easier and more transparent.

This external profit is not what the EU ETS was designed for. These financial companies are not reducing their emissions or helping others reduce theirs any quicker. Their playing of the market is unsurprising given they are profit-seeking businesses, but this entirely self-interested approach is not good here. They have developed a vested interest in the EU ETS, not just on the carbon trading side but also in the CDM offsetting scheme. Many of these companies engage either directly in carbon offsetting, or own firms that do this, allowing them to act on both sides of the carbon market. For example, JPMorganChase owns EcoSecurities and ClimateCare, while Goldman Sachs is a strategic investor in BlueSource.[12] Clearly, this creates a massive conflict of interest: the higher the price of carbon allowances, the more demand there will be for carbon offset credits. These firms therefore have an even more significant incentive to use their size to buy carbon credits and push up their price, exploiting and undermining the EU ETS as a business opportunity.

Large carbon traders, who have considerable influence in the running of the EU ETS, revealed their unsurprising prioritisation of profits over carbon reductions when they lobbied against the banning of HFC-23 from the CER market; and while the ban has gone ahead, they were successful in securing a delay in its starting date (see p.24). Discussing the industrial gas credit trade, the *New York Times* used the example of a Chinese refrigerant plant that was installing an incinerator at a cost of $5 million to eliminate HFC-23 and would receive $500 million from CER credits. It explained:

> The huge profits from that will be divided by the chemical factory's owners, a Chinese government energy fund, and the consultants and bankers who put together the deal from a mansion in the wealthy Mayfair district of London.[13]

Even the World Bank has a less than transparent stand on industrial gas credits and is involved in their trade. It has claimed that there was not enough evidence to suggest that the CDM was being exploited and that more of the gas was being produced in order to generate credits. According to its website, the World Bank had investments in two major Chinese HFC-23 schemes via its Umbrella Carbon Facility and bought '130 million CERs through 2013' worth around €1.6 billion.[14] When challenged about this, and its defence of HFC-23's inclusion in the CDM, it: 'cited rapid economic development as a reason for the swift growth in HCFC-22 [of which HFC-23 is a by-product] production in emerging nations, which it put at 25 per cent per year.'[15] The conflict of interest as an investor is keenly apparent, and one which such an organisation should not incur. If even the World Bank is manipulating the CDM and ETS, it is clear there is little hope for ensuring that private companies act properly.

> *...huge profits...will be divided by the chemical factory's owners, a Chinese government energy fund, and the consultants and bankers who put together the deal from a mansion in the wealthy Mayfair district of London*

Criminal damage

The EU ETS carbon market is large and complex enough for fraudsters to take advantage of it and disappear without a trace. Given that the whole system, including the EUA and CER certificates are electronic, it is very hard to prevent this from happening. Indeed, the European Law Agency has estimated that 90 per cent of all market volume in 2009 was caused by fraudulent activities.[16] This high volume of crime is made possible by the dramatically varying rules governing carbon trading that vary from country to country, meaning that only half of Europe's registries actually have any form of security. Some, especially in Eastern Europe, have very weak requirements, so false companies are easy to set up.

> *...the European Law Agency has estimated that 90 per cent of all market volume in 2009 was caused by fraudulent activities*

Below is a brief history of major frauds in the EU ETS:[17]

- January 2009 – The widespread phishing (explained below) attacks on users of EU ETS registries prompts the EU to revise Internet security guidelines.
- September 2009 – European Commission proposes measures for a consistent response to deal with VAT or carousel fraud detected in the market in 2009-10

- March 2010 – Hungary sells CERs that had already been surrendered to it under the EU's emissions trading system. In response, the EU amends the registry regulations to prevent CER recycling.
- November 2010 – Unauthorised access to EU ETS registry accounts in Romania results in the theft of 1.6 million EUAs.
- November 2010 – German Registry closes due to Trojan virus Nimkey.
- January 2011 – Discovery of an EU ETS-wide theft of €30 million worth of EU allowances leads to the closure of national carbon registries, the suspension of spot trade, and the implementation of an EU-wide upgrade of registry security.
- Money laundering is also alleged to exist within the European carbon markets, although this has yet to be confirmed.[18]

The latest and largest scandal was the one that took place in January 2011, primarily in the Czech Republic which alone lost seven million credits. The whole crime was made possible through 'phishing', where the criminals obtained the passwords to ETS accounts and then transferred the credits to themselves in order to sell them on. As a result, all ETS transactions were suspended for over two weeks as all national registries were closed and slowly reopened, with the Czech registry closed for two months afterwards.[19] The consequences are still reverberating. The stolen credits passed into many national registries, with the UK thought to hold 500,000 of these credits. This eroded the confidence of many traders because if it was found that they were in possession of the false credits, even without knowing it, these would be confiscated. Given only law enforcement agencies are able to tell the difference between genuine and black-market EUAs, this suspicion devalues all credits and the system as a whole.

The ability to defraud the ETS is less due to the EU's own security system than the individual countries' registries, which have varying levels of protection. The mere fact that there are so many registries has opened the gate to a form of 'carousel fraud', which has cost the EU ETS €5 billion so far, in 11 countries.[20] This involves the importation of EUAs, tax-free, from other countries and then charging the eventual buyer a VAT tax, which is then never paid to the relevant government. This additional false levy is simple to add on and hard to trace once the transaction is complete. While the EU has

The Italian Power Exchange was so overwhelmed by VAT fraud in 2010 that it had to suspend all carbon trading after an estimated €500 million in VAT was lost

been trying to improve its detection and prevention of carousel fraud, the same cannot be said for the national registries, which continue to be the security-related Achilles' Heel of the scheme. The Italian Power Exchange was so overwhelmed by VAT fraud in 2010 that it had to suspend all carbon trading after an estimated €500 million in VAT was lost.[21]

The whole criminal process is made much easier thanks to the ability to recycle used CDM credits, a process that has only recently been ended in 2010. Until then, it was possible for companies simultaneously to sell these on both the commercial and voluntary markets, supplying the same credits to those needing to buy more allowances, and enabling those buying them to enhance their green credentials. This double selling could net the original buyer a tidy profit. Other firms cut out the market and used the CDM credits as required allowances, then reused them to create the image of further, voluntary reductions. Both of these tactics further undermined the already shaky reputation of CDM credits: the suppliers have already been seen to act suspiciously, and now so do the buyers.

Additionally, once credits have been used as an allowance to emit CO_2, they are cancelled, but some governments and companies happily sell these ex-credits for extra revenue. Firms often buy these null credits as a display of their green credentials. However, unscrupulous organisations can buy the dud credits and pass them off as new ones, re-entering them into the carbon market. What happens next depends on the registry these were passed through. National jurisdictions may or may not allow the credits to be used anyway, giving some buyers an unfair advantage and reducing criminality to mere gambling. In those where the stolen credits are rendered useless by the law if discovered, it is the end company that loses out when it finds it bought fenced goods. The wider impact is fairly self-evident. Once the market realises that credits circulating might not be unused, the price subsequently collapses. This was seen in the Hungary example discussed above (see p.XX). The second-hand CERs forced the French and Nordic exchanges to close, while the price of the credits dropped from €12 to €1.[22]

Clearly, there is a need for more security than is currently the case. That companies have lost millions of credits from their carbon trading accounts through phishing scams just asking for their passwords shows how weak it currently is. Moreover, the need for precautions will only rise with time. Auctioning is still a relatively marginal practice compared to the allocation of credits, but during Phase III, the trend will reverse. If the crime levels are not brought into check, the scheme will be brought to a standstill. Indeed, the issues are so basic that the president of the International Emissions Trading Association wrote an open letter urging reform:

> [We] believe that basic security precautions, identity checks and active regulation, all familiar from financial crime prevention in other markets, could resolve the problems at once.[23]

While it is rather ambitious to assume basic measures would act as a panacea to cure the System of crime, there is a kernel of truth in his words. In response to the pressure, the European Commission has mooted plans to create a single European wide registry, which, if it is accepted by member states, will come into force in late 2011 at the earliest. This will replace the Community Independent Transaction Log (CITL) database with the new European Union Transaction Log. UK registration with this is compulsory. However, the suggested improvements, of a 24-hour delay on registry transfers, a minimum of two people authorising permit transfers and a ban on criminals convicted in the last five months of financial crimes from having accounts, are not enough. In all likelihood, these will create red tape for the law-abiding account holders, while the criminals find ways round and continue to wreak havoc on the scheme. In addition, not all registries will be covered. There are many outside Europe who buy and sell EU credits, which could not be forced to adopt the common platform. If they employ sub-standard security, then the scheme will be externally vulnerable. Passive laws are inadequate: an active watchdog is required if the ETS is going to continue. Moreover, with some countries opting out of the common auction platform, there is a need to ensure that every registry operates with the same level of security, but this cannot be enforced. There is little point in Britain investing heavily in security, if other states allow their registries to become corrupted and act as a gateway to infect all auction platforms.

It would be far more sensible to create some sort of regulatory body to monitor and protect the EU ETS, and it should take a form akin to the FSA. In addition, it would be eminently sensible to check the background of any company applying to join in carbon trading. Not only would this stop fraud, but it could also end the exploitation of the market by companies with conflicts of interest. Indeed, perhaps the best solution would be to cut out the middlemen altogether and only allow the 12,000 firms *needing* carbon credits to trade them. The EU ETS was not set up so that corporate banks could profit from it, but to help the environment at the lowest economic cost.

The EU ETS was not set up so that corporate banks could profit from it, but to help the environment at the lowest economic cost

Government revenue

The British Government is estimated to receive between €4 billion and €8 billion per annum from EU ETS auction revenues during Phase III.[24] This is a huge sum, the equivalent of between 0.8 per cent and 1.6 per cent of all UK tax revenues from 2010-11, but even this pales in comparison with the likely overall revenues reaped by the EU ETS through auction proceeds: €50 billion by 2020.[25] However, in the UK, these spoils are not necessarily going to be spent on low-carbon projects, contrary to the recommendations of the Stern Review and common sense.

As with other EU ETS members, the UK government has accepted a non-binding declaration that it is willing to spend at least half the revenue on climate-change related solutions. The official wording of the ETS Directive describes it thus:

> Those revenues should be used to tackle climate change in the EU and third countries, *inter alia*, to reduce greenhouse gas emissions, to adapt to the impacts of climate change in the EU and third countries, especially developing countries, to fund research and development for mitigation and adaptation, including in particular in the fields of aeronautics and air transport, to reduce emissions through low-emission transport and to cover the cost of administering the Community scheme. The proceeds of auctioning should also be used to fund contributions to the Global Energy Efficiency and Renewable Energy Fund.[26]

However, the British government disagreed with the principle of this and DECC has noted:

> On hypothecation (also known as earmarking), the Government is opposed to the Commission's proposal that a proportion of auctioning revenue should be dedicated to climate change measures. Whilst earmarking clearly appeals to many stakeholders, it is an inefficient means of determining public expenditure priorities.[27]

Having signed up to the agreement, it is likely Britain will therefore be making the most of its non-binding nature and ignore it. Indeed, the government's own EU ETS auctioning website admits in small print that: 'The money goes into the Government's consolidated fund for general spending purposes.'[28] While the Treasury might not like the idea of hypothecating in general, this is an area where an exception to the usual rule must apply. Despite the intentions of the ETS, from a British perspective, the current approach makes it appear more like a stealth tax than an environmental tool, regardless of the motive behind the scheme and even more so given VAT is payable on credits purchased at auction. From the UK's perspective the scheme appears to be benefiting the Treasury more than society and the environment. This lack of certainty undermines investor confidence as well, given income is not being directed towards clear projects and ends.

If the government feels that emissions are such a significant problem that financial sanctions are required, then it should be willing to play its part by using the resulting revenue to help provide a solution. Hypothecation should be limited to investment in low-carbon energy research and alleviating fuel poverty – an inevitable and unwelcome side effect of the EU ETS (especially given the considerable sum generated, which, in the grand scheme of government income, represents a tiny fraction of overall revenue). Moreover, on an EU wide scale, the agreement to spend funds on curbing emissions does not specify in any greater detail how this should be done, and recommends that some of the funds be used to help developing countries reduce their emissions, a situation likely to fall into the same pitfalls as the CDM. It would be far more useful to inject the funds directly into low-carbon energy R&D, but there is no mechanism to ensure this happens.

CHAPTER 4

The International Perspective

- The EU ETS is likely to cause carbon leakage, with a net rise in global emissions and a loss of thousands of jobs in the UK alone.
- Allocating 100 per cent of allowances for free to energy-intensive industries in Phase III is not enough to keep them in the UK due to the myriad of other environmental costs they must pay.
- The average energy-intensive company most vulnerable to carbon leakage currently faces environmental levies on its energy bill of £3 million, but, by 2020, this will be over £17.5 million. Of this, the EU ETS cost is currently negligible, but by 2020 will reach £3 million.
- In the UK, while *production* of carbon has fallen by about 15 per cent between 1990 and 2005, once the carbon imports are included, carbon *consumption* has actually gone up by around 19 per cent in the same period.
- Aviation is supposed to be included in the EU ETS from 2012 and this will erode European airline profits, potentially by as much as €40 billion by 2022. However, the details of the workings of the scheme are not known, so while airline tickets for 2012 are already on sale, companies can only guess at what the effect of the EU ETS will be.
- Extra-EU airlines are being included if they fly into Europe, which had led to China, the US and Russia taking the EU to court and threatening retaliation and trade wars. If they are included, international flights could avoid Europe, using extra-EU hubs and creating further emissions.
- Many other countries, when deciding how to reduce their CO_2 emissions, looked at the cap-and-trade model and rejected it or significantly watered it down. They were not prepared to undermine their economies. The ETS goal of linking to other schemes has failed.

Increasing carbon leakage

Windfall profits are an unwanted side-effect of the EU ETS, and while companies might be cashing in, this is due to the scheme's weak and inefficient nature. This wastefulness should be condemned. However, this does not necessarily mean the same firms could withstand full exposure to the ETS: a balance must be found.

It is important to highlight that there is a difference between carbon leakage and competitive concerns. While all carbon leakage is a worry from an economic and employment perspective, it might be the case that there is an environmental benefit, if the flight involves relocating from a dirty European factory to a clean one elsewhere. However, there is a further complication. While many of the émigrés will be moving to modern, energy-efficient plants, some of these factories will be powered by the most carbon-intensive fossil fuels. Johnson Matthey has developed a new catalyst called APICO for chemical production, which delivers shorter start-up times, has a longer life and generates fewer by-products. Most of the customers for this energy efficiency product, though, are in China, India and Brazil where it is used in coal-fuelled plants.[1] In this way, while the energy costs are kept low for the manufacturer, the environment does not benefit from the modernisation, and the UK continues to suffer in competitive terms.

It has been claimed by some environmental groups that the ETS will only have a marginal negative effect on the competitive edge of the UK economy:

> Significant impacts from emissions trading on the competitiveness of the UK economy have only occurred in a small proportion of industrial activity that is worth around one per cent of total UK GVA at a €15/t/CO_2 price.[2]

This plays down the problem: one per cent of UK's economy (around £15 billion and 290,000 jobs) is not a trivial sum and, even then, this is an underrepresentation of the reality. Almost all energy-intensive industries will, by their very nature, be liable to pay higher costs than their extra-EU competitors and their value to the economy has also been downplayed: the chemical industry alone accounts for 1.5 per cent of the UK's GDP and would be highly vulnerable to credit price rises.[3] As the only positive contributor to the UK trade in goods balance, and an employer of 600,000, its decline would be devastating for the country.[4]

Much criticism has been levelled at the fact that many energy-intensive industries receive free allocations which have developed into a surplus. Figure 4:1 (see p.41) shows the reason why, using the example of the cement industry, which has been particularly criticised. Initially, during Phase I, the typical over-allocation of the ETS's underwhelming targets occurred. During Phase II, it can be seen that, while the allocated allowances barely shifts, the recession takes its toll and devastates production of cement, hence a dramatic downward shift in emissions. These companies are still being allocated allowances on a business-as-usual basis, hence the continued over-allocation. However, had the recession not occurred, this would be a very different story, with business-as-usual production forced to take the EU ETS into account: in 2008 it would have been likely that emissions exceeded allocations, forcing companies either to reduce their emissions or, more likely, purchase extra credits. Once pre-recession levels of production resume, around the same time that Phase III kicks in, the return to higher emissions will eat away at the credit surplus rapidly.

Figure 4:1

The effect of the recession on the cement, clinker and lime sector throughout the EU-25

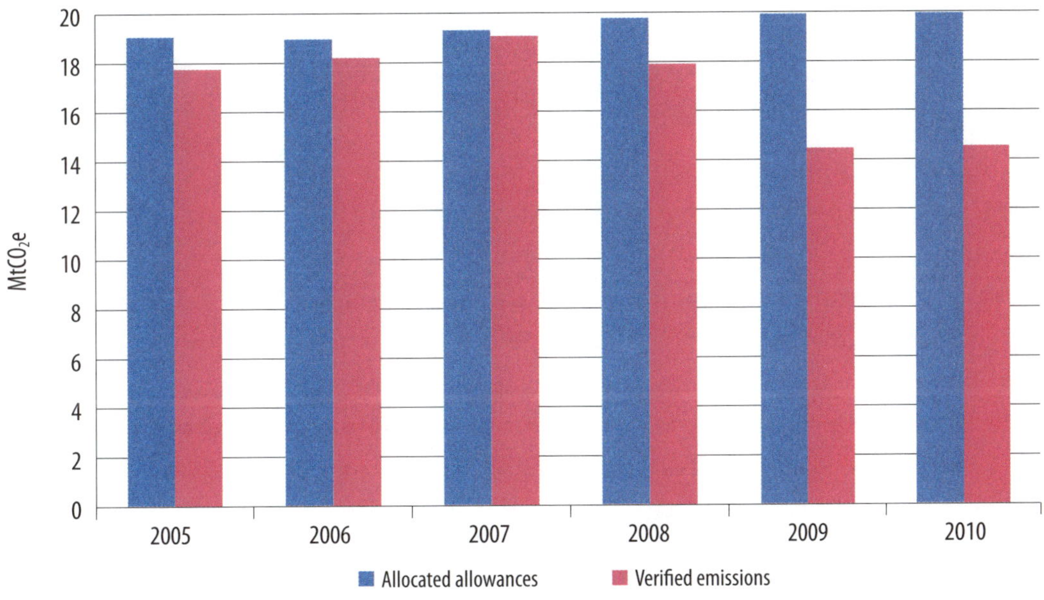

Source: European Union Emissions Trading System (EU ETS) data viewer

The important point is that, in the UK, the EU ETS is only a fraction of the total green costs payable by industry: the Climate Change Levy is also attempting to reduce emissions, the new carbon price floor will take this further and the Renewables Obligation is focused on incentivising renewable power generation. Many businesses pay more green levies than these, and in total they are overwhelming. Bills reflect this and a study of their effect has been conducted by the consultancy Waters Wye Associates.[5] For an average energy-intensive company, of the sort most vulnerable to carbon leakage, the additional cost of environmental levies on their energy bill is currently £3 million, but, by 2020, this will be over £17.5 million. Of this, the EU ETS cost is currently negligible, but by 2020 will cost £3 million (excluding the higher costs they will pay for energy, and other green charges as well).[6] It is this cumulative effect that the EU ETS contributes to that is highly damaging and for this reason free EUA allowances are absolutely necessary. Of course, there is a substantial difference between negating the cost of the EU ETS and providing windfall profits: only the former is acceptable, but carbon leakage is a real threat to be avoided.

However, not only is the increasing cost of all of these heightening the risk of carbon leakage, but it is also undermining the effectiveness of each individual policy. Even the Environmental Audit Committee had trouble working out quite where the EU ETS was having an effect:

> There is evidence that participating firms are incorporating the EU ETS into their business decisions, and adopting some measures to improve their carbon efficiency, though it is difficult to isolate the precise extent to which the EU ETS is a decisive factor.[7]

Indeed, this is why some companies vulnerable to carbon costs have continued to leave the country, despite being offered free EUAs: the EU ETS is only part of the issue. For example, the Teesside Corus plant was mothballed in December 2009 with a loss of 1,700 jobs, despite receiving a continual free allocation of permits to the tune of £250 million over three years following the mothballing.[8] The only silver lining here is that a Thai company bought the plant, although it is hiring only 800 employees to work there.[9] The worse the total costs become, the less likely it will be that new investors will be attracted to the UK: Britain will not just lose what it has already got, but will also not get what it might otherwise have got.

Hence, contrary to claims, carbon leakage has already happened and noticeably to the extent that the EU's GHG balance sheet has seen a significant reduction in emissions as a result. This reduction has been interpreted in political and environmental circles as proof that low-carbon investment has been taking place, but this is highly misleading. Actual emissions are merely being exported outside Europe as companies migrate and therefore the GHGs now exist on someone else's books. However, this does not mean Europe can turn a blind eye (if it was as serious about reducing global emissions as it claims to be, it would never have used these underhand methods anyway) and the EU still has a responsibility. Many of the products of émigré companies are imported back into the EU, causing further emissions in the process and compounding the negative effect. That this is currently ignored is due to the design of the EU ETS, which deals with carbon production only, not carbon consumption. Dieter Helm, Professor of Energy Policy at Oxford University, has stated:

> In the UK case, though the *production* of carbon has fallen by about 15 per cent between 1990 and 2005, once the carbon imports are added back (and aviation and shipping taken into account too) carbon *consumption* has actually gone up by around 19 per cent in this period.[10]

The economic sacrifices being made are therefore pointless, and emissions are effectively being outsourced. The EU is just as carbon dependent as ever, but is finding ways to hide this fact, via the EU ETS. There is simply no benefit, economic or environmental, to losing these industries.

The alternative to allowing this emigration, the protection of at-risk sectors from the otherwise inevitable carbon leakage, is all very well, but highlights the overall weakness of the EU ETS. Protected firms would have no motivation to curb their own emissions as no financial incentive would exist. A proposed alternative is 'border tax adjustments', so that importing products from non-EU ETS countries brings the price in line with products made by EU ETS installations.[11] This only solves half the problem. Given many energy-intensive industries rely on exports for their revenue, they would still be competing against cheaper rivals. Again, a solution of export rebates has been provided. This system would be highly complicated and add many layers of rules to the existing quagmire of corporate regulation. Companies that import affected extra-EU raw goods and export finished ones back out would be handed a red tape nightmare to deal with. In addition, there would be a large public cost, with the need to monitor companies on both sides of the deal. Then further questions arise: are there differing import tariffs for low-carbon suppliers and how can this be scrutinised? If low-carbon imports are not given a reduced tariff, then there would be no incentive for extra-EU firms to reduce their emissions, so a form of sliding scale would be required. This would then require some sort of third-party monitoring and regulation, but this could easily lead to a CDM-esque situation that values profit above emission reductions – an unacceptable outcome. In short, the idea, while theoretically desirable, would be very difficult to impose upon the real world without burdening industry further.

Aviation and the EU ETS

As was stated earlier, from 2012, aviation will be included in the EU ETS with a cap in 2012 of 97 per cent of 2004-06 emissions and 95 per cent from 2013.[12] This is due to the perceived contribution of the industry to carbon emissions. It accounts for something like three per cent of European CO_2 output, and the likelihood of passenger numbers rising by 2050 will result in emission increases at the time when other sectors' are falling. The UK accounted for 24 per cent of the EU-15's total CO_2 emissions from aviation, and its inclusion has been strongly supported by Britain, which made it a goal of the UK's EU Presidency in 2005. However, the introduction has been delayed for many years due to its controversial nature, as it would apply not just to EU airlines, but all extra-EU ones that fly into Europe as well.[13] Unsurprisingly, this has caused much anger and has had its legality challenged. At best, this will lead to red tape and increased costs to consumers. At worst, the policy will spark an international trade war with ugly consequences.

The underlying assumption of its inclusion is that any extra cost can, as in the case of some other industries, be passed on to the consumer. Secondly, it is thought that the price of flights will not affect their demand.[14] However, the cost of a flight ticket could rise by up to £34 per passenger, per flight, according to Standard & Poor's, which will have a significant impact on many families' ability to fly. Standard & Poor's has estimated that, at the current price of around €15 per tonne of CO_2, in 2012 alone the aviation industry will have to pay €1.125 billion.[15] In total, Ernst & Young has suggested that the EU ETS will erode European airline profits in the long-term: potentially by as much as €40 billion by 2022.[16] As with energy-intensive sectors, there is a risk that the ETS will undermine the viability of a European aviation industry. Similarly, this is hoped to be mitigated by providing free permit allowances which will make up 85 per cent of the total, until at least 2020. Just as elsewhere, this could generate windfall profits as seen in the power sector, with the non-existent costs still passed on to consumers, a highly undesirable outcome.

Already, the ETS is having a negative effect on the aviation industry because, while the inclusion of airlines is now less than a year away, there is still a huge level of uncertainty surrounding the scheme, not least because not all member states have actually formally agreed to the plans. The ratifying EU Directive 2008/101 was supposed to have been made national law by each state by February 2010, but this deadline failed to account for the complexity of adopting the legislation – the first of its kind worldwide. Germany took until July 2011 and three other countries took even longer.[17] All the deadlines associated with aviation's inclusion have been missed, making the transition highly disruptive to airlines. No delay was allowed for in timing aviation's inclusion in order to take into account the sluggish political process, meaning airline tickets for 2012 are now already on sale, but companies can only guess at what the effect of the EU ETS will be.

...tickets for 2012 are now already on sale, but companies can only guess at what the effect of the EU ETS will be

The 'saving grace' of aviation's inclusion is that it universally applies to all airlines that fly in and out of Europe, regardless of whether they are from EU member-states or not. The rationale behind this is clear and the EU's own transport ministers warned: 'in order to maintain the Union's competitiveness, similar commitments should be sought at international level'.[18] In theory, this will occur, meaning an equality of cost and therefore competitive edges will not be blunted. In practice, this is a legal time-bomb and many countries have said they are prepared to take action if the EU presses ahead with these plans. Most worryingly, a coalition of giants has emerged, with the US, China and Russia all publicly stating their opposition to being forcibly included and their willingness to take action to prevent this. The chairman of the Chinese Aviation Transport Association stated that China is 'ready to sue the EU at any time... all the Chinese airlines are against this plan – it is not legally binding and it is only useful in Europe'.[19] The latter phrase concisely sums up the whole problem with the EU ETS: the rest of the world is not prepared to weigh in. The Air Transport Association of America (ATA) (now renamed Airlines for America), the industry's US trade body, has gone further and applied to the European Court of Justice to rule that the non-EU inclusion is illegal, citing that transatlantic flights have little impact on European emissions. As evidence to the ECJ, it cited that on a flight from San Francisco to London Heathrow, only nine per cent of emissions are actually released in European airspace and the other 91 per cent are none of its business.[20] The ATA has advocated an alternative, global scheme whereby the entire sector regulates emissions without unilateral action by region or country. The ECJ is not expected to make a decision until the beginning of 2012 at the very earliest, by which time EU airlines will already be paying.

The chairman of the Chinese Aviation Transport Association stated that China is 'ready to sue the EU at any time

If these legal challenges succeed, the consequences for the European airline industry will be dire, with uneven costs having a magnified effect in what is a very competitive market. Similarly, if the challenge fails, the results are equally undesirable. The risk of universally forcing compliance to the EU ETS is the backlash from airlines unwilling to

The risk of universally forcing compliance to the EU ETS is the backlash from airlines unwilling to pay and therefore suspending flights into Europe

pay and therefore suspending flights into Europe. The economic consequences of this would be magnified, not just in the lost revenue of airport slots, but fewer business and tourist trips, costing the economy valuable missed opportunities. Carbon leakage is also likely to occur. Passengers unwilling to pay the added fees will simply use carriers with extra-EU transport hubs. Depending on where these are and what the journey is, this could lead to substantial and unnecessary extra emissions.

Willie Walsh, the head of British Airways and Iberia (BA has estimated it will face annual costs of €50 million)[21] has argued the effect will be even greater. At the 2011 AGM of the International Air Transport Association he said: 'It is clear that the countries are going to retaliate, whether in the form of imposing additional taxes on European airlines or restricting access to markets.'[22] Walsh is not alone in fearing this, and Airbus, Virgin Atlantic and others have all voiced concerns as well. This is the worst possible outcome, as it would undermine the equality of cost that the global implementation of the EU ETS is supposed to deliver, but so too would an inter-Europe scheme, making this a lose-lose situation. The threat of a trade war is real, and it is not a battle Europe would win.

Even if the EU presses ahead with aviation's inclusion, the same bugbears undermining the wider EU ETS will also affect it and the targets will be met through permit purchases, rather than actual emission reductions by airlines. Even the official parliamentary briefing paper on the subject highlights this, on its first page:

> It is expected that the majority of the cuts will be met by airlines purchasing international credits created through the Kyoto protocols rather than through the purchase of EU ETS credits or reducing their own emissions.[23]

The European Commission defended this:

> Providing aviation with these options does not reduce the environmental impact of the proposal since the climate impact of emission reductions is the same regardless of where they are made.[24]

The Commission claimed the inclusion of aviation would see emission reductions of 183 million tonnes of CO_2, a 46 per cent reduction on 'business as usual levels' as emissions will be capped at 2004-06 levels. This theoretical level is substantially higher than the actual effect if the CDM system is being used. The parliamentary briefing paper comments: 'other options are available to the aviation industry and actual cuts are not likely to be anything as significant'.[25] In terms of revenue, the British government will receive a significant cut of whatever money is raised through auctioning permits. Out of 4,000 European airlines, 891 of these have been allocated to the UK to regulate and consequently receive revenue from, a number second only to France. However, as has already been discussed, it is highly unlikely that the government will spend the revenues purely on environmental projects.[26]

The international rejection of the cap-and-trade model

While the EU ETS might be the largest carbon trading scheme in the world, accounting for 84 per cent of the global carbon market, it is not the only one. It was designed to link up with other schemes in time, to create a worldwide cap-and-trade network, and at the time of its inception, there was a great deal of optimism about what the future held for carbon trading, especially in the global economic powers. The forebears and cousins of the EU ETS, spawned around the globe, have not fared much better than the dominant system and, as a whole, the World Bank has warned that the international carbon market has stalled since 2010.[27]

Many countries have decided against introducing cap-and-trade schemes, undermining the positive goal of eventually creating a global emissions trading system and continuing to leave the EU exposed to the economic difficulties the EU ETS has created. Below are a few brief summaries of other countries' experience with these schemes:

- Australia: The level of passion in Australia over a mooted ETS has been intense to say the least. The Carbon Pollution Reduction Scheme was created in 2008 and due to be introduced in 2010 but party politics got in the way. Permits were estimated to cost between $20AUD-$40AUD (£13-£26) and would have affected less than 1,000 businesses. The bill was criticised for by some for being unambitious and by others for being economically damaging. After various modifications, including delayed starts and more free allocations, the scheme was delayed in April 2010 by then Prime Minister Kevin Rudd until after the Kyoto Protocol expired in 2012. This decision was justified by citing the lack of international clarity and commitment to reducing carbon emissions, especially from the US, India and China. The ETS idea has been abandoned and a carbon tax has passed into law instead in November 2011. This will affect 500 companies and the opposition has pledged to repeal it if it comes to power.

- China: The largest carbon-emitting country in the world has a unique approach towards adopting an ETS. Policy makers have found consensus in the idea that an ETS-style system will reduce carbon *intensity* by 40-45 per cent by 2020, rather than carbon *emissions*. This is due to Chinese unwillingness to overburden their industries by imposing targets that could jeopardise their competitive advantage over international rivals. Instead, given the carbon-intensive nature of many Chinese firms, where unit for unit they produce more carbon than Western counterparts, the room for improvement in carbon efficiency will be focused on. The ETS will be designed to allow a growth in carbon emissions and credits will be earned from meeting energy intensity targets. There are no plans to link the scheme internationally, which is unsurprising given its unusual and somewhat incompatible nature.

  So far, six regions have been selected for a pilot ETS to be rolled out by 2013 and a national system is planned to begin before 2015. However China has faced problems in the past with energy efficiency campaigns. During the eleventh five year plan (2006-10), officials cut electrical supplies to thousands of businesses in a last minute attempt to meet emissions and pollution targets.[28] It remains to be seen how the ETS goes ahead, if at all.

- South Korea: Initially, the plan was to start emission trading in 2013, but the pressure from industry meant this was delayed in February 2011 until sometime before 2015 at the latest. Industry leaders had been complaining that, as rival countries like Japan or the US had decided against similar schemes, the competitive impact would be too great. In addition, more allowances will now be allocated rather than auctioned, up from the 90 per cent pre-delay to somewhere between 90 and 100 per cent, and there will be lighter fines for non-compliance.[29]

- Japan: In March 2010, the 'Basic Act on Global Warming Countermeasures', which included an ETS, was put forward to be voted on by the Diet. While mostly accepted, the ETS idea was heavily criticised as too costly to the national economy and incensed industrial groups, with the scheme theoretically to come into force in 2013. Given the resistance, the ETS section of the Act has been delayed and will be discussed sometime in 2014, a decision augmented by the reluctance of other countries to impose similar schemes. With a carbon tax beginning in October 2011, it is unlikely the Japanese ETS will ever go ahead.

- The United States: A variant of the EU ETS would have been set up in the US by the American Clean Energy and Security Act of 2009 but it was blocked by the Senate in July 2010. The proposed scheme would have run from 2012 to 2050, to deliver 17 per cent emission reductions on 1990 levels by 2020, with caps set for each year. Initially, 85 per cent of allowances would have been allocated, and revenues from auctioning were to be used to subsidise low-carbon energy projects.

  Independently of the federal scheme, California, the world's ninth largest economy by GDP, has a cap-and-trade system, which has been passed into law. However, legal challenges delayed its implementation from January 2011 to 2013 at the earliest. In March 2011, a court ruled that alternatives to the ETS had not been sufficiently considered, and an appeal is currently in progress. The State is allowed to continue to develop the scheme while the matter is deliberated.[30]

The key theme throughout these different experiences has been the unwillingness to commit if it appears no other major power is doing so... all the above nations are top global economies and want to stay that way

The key theme throughout these different experiences has been the unwillingness to commit if it appears no other major power is doing so. This is very significant given that all the above nations are top global economies and want to stay that way. As none of them have been willing to make the first move (second, if the EU ETS is included), an international stalemate has emerged where industries can lobby and somewhat justifiably claim that the economic consequences of unilateral cap-and-trade schemes will be harmful. In a survey by the World Bank, it was found that very few respondents felt there would be 'a new legally-binding multilateral framework... with legally-binding commitments to reduce emissions'.[31] The survey revealed pessimistic responses until 2020, and only after then do people believe a new framework will emerge. For Europe to go it alone for eight years, and increasingly step up its efforts, while the rest of the world sits and watches, is futile: an alternative, globally beneficial way must be found. Much support for the ETS was grounded on the basis of a global cap-and-trade scheme. Given that this has failed to materialise, its usefulness is severely limited.

CHAPTER 5

Underestimated Complication and Political Opportunism

- The EU ETS is the jewel in the crown of EU environmental policy, and this special status means the European Commission has been unwilling to tackle its major problems head on.
- The EU ETS is not the most economic method to reduce emissions. It is far more complicated than the simple principle of cap-and-trade would suggest.
- Despite Phase III introducing a common auction platform, many countries (including the UK) have opted out of this, so the security and coherence of the scheme will continue to be undermined.

The EU ETS was not properly thought through when it was designed. The principle of a cap-and-trade scheme is simple enough to understand on paper, but in reality the mechanisms for trading and the confusing relationship between industries, power generators and their emissions means the complication has been underestimated. The scheme relies on very optimistic assessments to justify its existence and much of the debate supporting it is based on sweeping statements with little or no evidence. The following is a typical example:

> The larger the part of the economy covered by the ETS, the more efficient will be the distribution of carbon across the economy as a whole. And the more efficient the distribution of carbon across the economy, the cheaper it will be to meet emissions reduction targets in the future.[1]

Whether by accident, or consciously to obscure the issues, the EU ETS's failures and the inability to overcome them are being overlooked by the political establishment, and many deep green environmental organisations would rather stick with the subpar scheme than acknowledge that alternatives might be better options.

The over-allocation of permits clearly shows that not enough was done to balance the negative impact of the EU ETS with meeting its goals. This was basically because the task was too huge for the EU Commission to handle, hence the use of NAPs in the first place. The countries were relied upon for unbiased accuracy, with the Commission effectively rubber-stamping their plans, unable to verify independently whether the requests were accurate. For this reason the difference between the submitted Phase II NAPs and accepted ones was just 10.5 per cent, only 1.95 per cent lower than the verified emissions of Phase I.[2] Evidently, the lack of scrutiny has led to many of the problems plaguing the Scheme and this is something that will continue, given that the breadth of the ETS will only increase. The EU cannot rely on the hope that, by Phase III, enough has been learnt to allow the scheme to run without any more hitches. In all likelihood, the bugbears will continue.

It is an illusion to suggest Phase III will be less complicated just because the NAPs will be replaced by a universal cap and auction system. The idea is that, from 2013, a common EU platform will be used to auction allowances, and the stated aim of this is to maximise efficiency, save money and reduce red tape, while minimising the potential for market abuse. However, at the same time, the EU also said that countries are allowed to create their own national platform if they so wish. This offer has been taken up by the UK, Germany and Poland, who opted out and will devise their own method. These auction dates will also deviate from those of the common platform's. The EC might have its own political reasons for wanting all countries to use its common method, but it does stand to reason that a myriad of platforms will create a new dimension of complication for the Scheme, making trading much harder and more expensive to monitor. Unless all these platforms have the same level of security, this undermines the reason for having a unified system. The only gains to be made from this are by the carbon exchanges, which can run both the auction platforms and will potentially be able to play one off against the other as fees and other costs will differ.

The economic basis for the ETS is less than convincing. It has already been seen that the Stern Report of 2007 was too optimistic and gave an unrealistically low cost of decarbonisation.[3] Total costs were downplayed so that mitigation was predicted to cost one per cent of global GDP per annum, but this relies on the cheapest forms of action taking place, and fails to leave margins for government failure and lobbying. Overall, it is judged that 'there is considerable evidence that the most expensive options are chosen first, not last' and the EU ETS forms a core part of the overly pricey European package.[4]

The oversights of the EU ETS are in part the result of it being constructed in the minds of politicians, and remaining a political tool. Dieter Helm, the Professor of Energy Policy at the University of Oxford, has argued the whole climate change agenda:

> ...provides the [European] Commission with a new project – following on from completing the internal market, monetary union, and enlargement after the collapse of the Soviet Union. Its citizens can find common cause in the new green agenda. The acrimony over the Lisbon Treaty and the referenda can be put behind it.[5]

The ETS is the jewel in the crown of this precious project, and as such retains a special place in the hearts of EU commissioners. The politicians might not be willing to contemplate its substitution by worthier emission-reducing schemes, but they are vulnerable to being swayed by lobbying to satisfy the needs and greed of certain industry interests. It is this that has led to the overly generous allocation of permits: politicians have nothing to gain by angering their supporters via raising business costs, nor by hindering their country's economy. This will continue in Phase III even without the NAPs. Lobbying will continue in a different form, as countries attempt to claim the lion's share of auction revenues to protect their own interests.

From national politicians to the EC itself, the only way to ensure that greater emission cuts are made is to remove the decision making from the hands of those who are susceptible to influence. The 'ideal' would be an independent body that would take the place of the EC. The body could also oversee the running of the credit markets, to prevent the fraud that so frequently occurs. However, such a body would also add to the level of bureaucracy that already exists to deal with the EU ETS, which is undesirable, and it is also very hard to ensure a body is truly independent. Ironically, the creation of the independent body would be vulnerable to the very politics it would seek to avoid: it would have to be brought into existence through a formal, legally binding treaty, and there can be no guarantee such a thing would be acceptable to all members, especially those who can influence the EC, while the timespan of this creation would be likely to drag on almost to the end of Phase III. The cost of creating and maintaining such a body is also a factor to take into consideration.

Scrapping the EU ETS would show the public that the politicians were wrong, and that costs of living have been raised unnecessarily. In addition, this would derail the gravy train benefiting so many people

While it was a good idea on paper, the EU ETS has failed to have the effect wanted. Scrapping the EU ETS would show the public that the politicians were wrong, and that costs of living have been raised unnecessarily. In addition, this would derail the gravy train benefiting so many people. Neither of these are desirable outcomes for the ETS's architects, and it is more convenient for them to keep the EU ETS operational, to save face.

PART TWO

Prognosis: The EU ETS is Incurable

CHAPTER 6

Ineffectively Propping up the EU ETS

- The faults of Phase II have not been dealt with and, as a result, Phase III will be ineffective for years after its inception.
- If the EU wants the large emissions reductions within the timeframe of the 2020s, then the failures cannot be allowed to continue.
- The EU ETS forces a lose-lose situation of offering installations free credits, and creating windfall profits, or subjecting them to auctioning, and risking their emigration to countries outside the EU.
- Rather than trying actively and positively to assist the transition from carbon-intensive to low-carbon energy, the EU ETS is designed to punish a 'business as usual' attitude. This is a passive stance, which has been seen to have little effect. Relying on market forces is not enough and the scheme has frequently undermined investor confidence.
- The UK Government effectively admitted the ETS had failed when it decided to implement a unilateral carbon price floor (CPF), setting a minimum cost for ETS credits. The revenues will not be earmarked for environmental ends, and household energy bills will rise by £6 in 2013 and £17 by 2016 at current prices as a result of it. The floor is effectively a poor imitation of a carbon tax.
- Up to 110,000 households could be depressed into fuel poverty by the CPF. It is also environmentally useless as it means that, via the ETS, other EU countries will be able to pollute with any emissions saved in UK.
- The EU's Energy Efficiency Directive, designed to shore up the ETS, will actually undermine it. By driving down energy consumption, fewer ETS credits will be needed and the price collapses.
- Investment funds designed to increase low-carbon investment alongside the ETS are too short-term and full of delays.

It appears to have been forgotten that the EU ETS is just an experiment, albeit one on a very large scale. No real emissions trading system had ever been attempted before, and there have been learning curves involved in setting it up and adjusting it. The lessons learnt from all of these are telling us that it is no longer a viable experiment.

As has been seen, the EU ETS is flawed, and in many ways far too deeply for recovery to be possible. The over-allocation will spill over into Phase III and windfall profits will continue, as will the profiteering of the middlemen, while the emission reductions it provides are tiny and often illusory. It has been allowed to progress without reform for too long and, as problems have manifested themselves, the likelihood of any revisions being able to fix these problems steadily decreases. While small revisions have been made to the ETS, the problem is that all the major corrections to the scheme, such as benchmarking, the 1.74 per cent annual cap reduction, 100 per cent power sector auctioning, and later revisions such as the removal of industrial gas credits, will not come into force until Phase III. This leaves over a year of inefficiency to go. As has already been discussed, the existing problems already mean that, regardless of the incoming fixes, many companies will not have to reduce their emissions until 2016-18. If the problems that remain in Phase III are similarly revised with delays in their implementation, then the System's effectiveness is further undermined.

There is an inherent contradiction in the EU ETS that cannot be resolved as long as the scheme operates. The energy-intensive sectors at risk of carbon leakage will be excluded from auctioning in Phase III to militate against their collapse. However, this will mean that some sectors will make windfall profits from the spare credit capacity. Of course, exposing them to significant auctioning could lead to their demise. Either way, the System is not able to deliver the optimum environmental or economic result. This inefficiency cannot be resolved, no matter how many revisions are made to it.

It is only now, with just over a year until Phase III begins, that the UK has launched an enquiry into its workings, questioning whether it will be useful in the future at all.[1] Now, the only changes possible within it would be compromises. Instead, the entire framework within which EU environmental policy works needs to be changed. Other methods must be investigated, to replace, or more accurately, supersede the ETS. Overall, the System has been hijacked, and become a route to extract as much money as possible out of good environmental intentions.

... the System has been hijacked, and become a route to extract as much money as possible out of good environmental intentions

The money to be raised from the EU ETS in Phase III once auctioning increases is impressive, with £4-8 billion generated per annum.[2] However, it must be remembered that this will not necessarily be spent on decarbonising the economy, as the government has already said that just as it refused to hypothecate Phase II revenue, the same will go for Phase III funds. On this basis, the usefulness of the ETS is much reduced for the UK. Also, as will be seen later, alternatives such as a carbon tax can raise larger sums (see p.93).

The EU deludes itself in thinking that the EU ETS is the only choice Europe has if it wants to reduce its emissions. This is simply untrue, and this conviction is only based on the fact that using another means suggests their flagship policy of market-based environmentalism has failed, which it clearly has. When criticism is levelled at the ETS, the response is an admission that all is not right in the scheme, but these are teething troubles that will be cured in time, if only Europe 'gives it a chance'. The ETS has run since 2005, with these inefficiencies still unsolved and emissions still not brought down. The European Commission cannot have its environmental cake and eat it: if it actually wants the large emissions reductions within the timeframe of the 2020s, then the failures cannot be allowed to continue. Waiting until late Phase III's relative improvement is not the best policy for the environment itself. The whole approach needs to be rethought, down to the very basics of what it is we need to reduce and where.

Arguments in favour of the EU ETS rely on the fact that it should reduce emissions in the long run, even if they barely fall, or even rise, in the short term. However, looking at the long run, there are far more attractive alternatives, which can offer larger reductions and with a smaller economic cost. If genuine emission reductions are the aim, rather than creating a means for a small elite to get rich(er), these alternatives should be seriously considered.

According to the Hartwell Paper, co-authored by many climate scientists and economists, any new policy would have to meet three criteria in order to be successful:

1. It should be politically attractive, meaning an approach which allows us to take a few small steps which offer rapid and demonstrable pay-back, thus helping to sustain the effort.
2. It should be politically inclusive, meaning an approach which is pluralist in instinct.
3. And it should be relentlessly pragmatic, meaning an approach which prizes progress that can be measured in the short as well as long terms.[3]

PriceWaterhouseCoopers (PwC) has identified a list of five concise factors which are also worth repeating:

1. Ability to deal with uncertainty about technology (and so the costs of abatement) and the environmental and economic benefits of abatement

2. Avoidance of undue carbon price volatility, which might weaken low-carbon investment incentives

3. Revenue-raising capabilities for government (including predictability of revenues as well as their expected levels

4. Simplicity [and related to this, administrative and set-up costs]

5. Political acceptability, both nationally and internationally[4]

Even on the most generous interpretation, the EU ETS fails to satisfy the Hartwell Paper's third criterion and PwC's second and fourth points. It only tackles the symptoms of the environmental problem, not the illness itself. Rather than trying to change the fundamentals of society and economies, to assist actively and positively the transition from carbon-intensive to low-carbon energy, the scheme is designed to punish a 'business as usual' attitude. This is a passive stance, which has been seen to have little effect. Relying on market forces is not enough, and leads to the market trying to take advantage of the situation, as is instinctive and therefore understandable if not commendable. Via the EU ETS, governments have tried to shy away from direct intervention, something that has clearly failed. It is time to reassess the value of this approach. If governments are prepared to voice their concerns about carbon emissions, then they should also be prepared to help solve the issue themselves.

The aim of environmental policy should be the decarbonisation of society at the lowest cost possible. This does not necessarily mean via renewable power, and the ETS makes no provision for renewables which are dealt with instead through the EU Directive on Renewable Energy. The approach of the ETS here at least is the correct one: non-renewable power is not in itself the problem, only its usually carbon-intensive nature.

Before progressing to this report's recommendation on what to replace the EU ETS with, it is worth examining the adjustments and alternatives already posited, which reveal how the EU ETS is no longer seen as the prime vehicle for change by the UK and even the EU.

The UK solution: a carbon price floor

Within Britain, the likely failure of the EU ETS to incentivise low-carbon investment has led to pressure for a carbon price floor. The Committee on Climate Change had originally forecast a carbon cost of cost of €56 per EUA by 2020, but has since revised this to €22 (CCC) and said:

> The carbon price is likely to be significantly lower in 2020 than we previously projected. This will have consequences for investments in low-carbon generation. A range of measures including tightening the EU ETS cap and a UK carbon price underpin should be seriously considered to strengthen incentives.[5]

As a consequence of the shortfall in cost, the government has also deemed the EU ETS inadequate, and the March 2011 Budget set out plans for a carbon price floor (CPF) with a minimum cost of €30 by 2020. The official document outlining the CPF justified it in these terms: 'for a variety of reasons, the carbon price has not been stable, certain or high enough to encourage sufficient investment in low-carbon electricity generation in the UK'.[6] The hope is that this will reduce the risks and speculative nature of the carbon market, as the lowest costs will be known and investors can act accordingly to deliver low-carbon power. The government is set to raise an extra £3.2 billion between 2013 and 2016 via the CPF. However, as with the EU ETS in general, the government is refusing to promise that the revenue will be spent on green issues. As such, it is a poor imitation of a real carbon tax. It is expected that household energy bills will rise by £6 in 2013 and £17 by 2016 at current prices as a result of the CPF.[7]

The principle behind a price floor is sound in theory. As Dieter Helm has argued:

> ...it is hard to think why one would not have a floor: what could the downside risk possibly be? For, if policy-makers genuinely thought that the carbon price might fall below the floor, there would be a credibility question about the scheme as a whole.
>
> Either the Commission believes that the EU ETS price will always be above the floor (in which case, there is no problem putting a floor in place), or it believes that the price could fall below (in which case, there is a good case for having a floor).[8]

A price floor would certainly improve the EU ETS, but *only* if it was implemented across the whole EU, or indeed the globe, to mitigate cost competitiveness issues. The real problem with this CPF is that it will be unilaterally implemented by the UK. This would mean the price of production and energy would increase in the UK above that of the rest of the EU, let alone the rest of the world, and the cost increase would be regardless of the other green levies Britain implements.

If Britain pushes industrial costs to levels above the rest of the EU, the results will be disastrous. Manufacturing, and particularly energy-intensive industries, already face

> ***Manufacturing, and particularly energy-intensive industries, already face very stiff competition from European rivals and costs play a critical role in determining the UK's competitive advantage. When the cost of British production is raised above parity, the result will be potentially widespread industrial collapse***

very stiff competition from European rivals and costs play a critical role in determining the UK's competitive advantage. When the cost of British production is raised above parity, the result will be potentially widespread industrial collapse. Many foreign-owned firms will leave the UK, either pulling out immediately or running British outlets purely for short-term profit and neglecting investment until maintaining production is no longer viable. Larger British-owned firms will act similarly, while those who cannot afford to leave will fold, buckling under the pressure of costs. Such a process may start off slowly, as costs are still manageable, but by the latter half of Phase III, when the effect of the price floor is likely to be significant, it is likely that industrial emigration will snowball, compounded by the dislocation of entire supply chains as firms move overseas. The potential damage this will do to the UK economy cannot be overemphasised. Looking at just one sector, the chemical industry, which is highly vulnerable to energy cost rises, 600,000 jobs are at risk if bills grow.[9] Along with the social consequences this loss will have, the industry has a turnover of £60 billion per annum and accounts for 15 per cent of UK exports, which would also be lost.[10] Lastly, and by no means least, the chemical industry is a vital supplier of low-carbon products, from insulation to fuel-efficient materials and without these, Britain's attempts at creating a low-carbon economy will be smothered in the cradle and low-carbon investments will not be made.

On top of this, the CPF does not mean UK emission levels are likely to fall. The Energy and Climate Change Select Committee has said:

> [S]everal witnesses felt that the EU ETS should be the primary instrument for putting a price on carbon because a UK-only carbon price will not result in additional carbon savings and could undermine the EU-wide system… There would be no net environmental benefit.[11]

Perhaps obviously, the CPF will fail to have any effect because it only applies to the UK. Given the nature of the ETS, where overall emissions are capped, any reduction in the UK's carbon output will mean that companies in a different country will be able to purchase the spare credits and emit that saved amount. Moreover, the emigration of industry will not automatically mean companies will rebase themselves in Europe. On the contrary, the forced move will mean many will take the opportunity to move to far more favourable business environments, where energy costs are minimal. Countries offering such deals – China, Russia and Saudi Arabia for instance – also have few, if any, emission regulations. This means carbon leakage is likely to occur, worsening the global level of emissions while the UK's own may well improve.

The only way for a price floor to work would be if it were implemented on a European level, a solution favoured by Lord Turner, chair of the CCC.[12] Even then, the carbon price floor undermines the original point of the System, i.e. delivering low-carbon investments at the lowest possible cost. Shoring up the scheme at an arbitrary level is more in line with the framework of a carbon tax, which is certainly not a bad method to use, but as the EU originally decided (see p.xxviii), it is an 'either/or' situation where either a cap-and-trade or tax system should be used, not a hotchpotch of both.

Even on a European scale, there is nothing to prevent firms passing the extra cost on to consumers, as they have been doing so far. In the case of a price floor in the power sector, the rise in electricity and heating bills would aggravate fuel poverty. The Government has already admitted this will be the case with the UK's own CPF: 'on its own, an increased wholesale electricity price would tend to increase the risk of fuel poverty for some households'. Indeed, the Government has estimated that, as a result of the price floor, up to 110,000 households could be depressed into fuel poverty.[13] The sacrifice of so many consumers for the sake of partially mending the EU ETS is unacceptable.

More widely, a price floor undermines the *raison d'être* of the EU ETS – its market-driven nature – which means it will fail to deliver the 'investor confidence' the government claims it can. Indeed, the only certainty it will give investors is that getting involved in the EU ETS is a risky manoeuvre: once the government has tinkered with the system once, it will be very difficult to guarantee that this is a one-off and no further price floors or adjustments will take place. If confidence in the scheme drops, then so will demand for credits, regardless of the price set.

The previous Labour government, aware of the flaws of the EU ETS, decided against using a price floor partially on this basis. Instead, it felt the best way to fix it was to reduce the number of credits available, and this led to the cap reduction of 1.74 per cent each year during Phase III. As has been seen, though, this is not really enough to spark the low-carbon revolution. If a market-driven solution is desired, then political needs cannot be allowed to trump economic ones. If it is felt that the market has failed to act according to the politicians' wishes, then an entirely new solution is needed, not just an adjustment.

Overall, the ECC Select Committee stated:

> We would have preferred the Government to establish a nominal Carbon Price Support level until 2018 and then set a long-term trajectory based on advice from the Committee on Climate Change. Until then, the Carbon Price Support represents little more than an additional energy tax, which will be passed on to consumers.
>
> Carbon Price Support is a short-term solution to the failure of the EU Emissions Trading System to deliver a meaningful carbon price. It poses risks to UK energy security and the UK economy more widely. The White Paper needs to justify its costs and benefits and provide a persuasive plan for its integration with the EU Emissions Trading System.[14]

This is a damning but unsurprising verdict. The CPF is a compromise between proving UK 'global climate change leadership' and trying not to overburden industries: it fails to achieve either end. Implementing the scheme on a European-level is possible, but does not fix all the problems of the EU ETS and so should be seen as a 'best of the rest' solution. It would be better entirely to replace the ETS so as not to continue aggravating fuel poverty but, if possible, to reduce it.

The EU solution: an energy efficiency drive

The EU is attempting to shore up its emissions reduction policy, not by scuttling or even repairing the leaky EU ETS, but by creating additional regulations. This is symptomatic of the EU's approach to the ETS: it has been burying its head in the sand about the scheme's failure, and it prefers to attempt to address the issues indirectly rather than tackle them head on, even at the risk of undermining the ETS further. As the flagship policy on climate change, the ETS might not be beyond mild reproach, but it does appear to be off-limits to restructuring. While other existing environmental regulations, such as the Industrial Emissions Directive, come close to covering the same ground as the ETS, they have so far fallen short of actually encroaching on its emission-reducing goal. However, as the inefficient nature of the ETS has come to light, additional policies are now being implemented to solve the same environmental problems, an approach that leads to confusion and downright contradiction. A draft Energy Efficiency Directive (EED) is in the process of being created and the very nature of such regulation suggests officials at the highest levels of the EU have little faith in the EU ETS to deliver. The proposed EED will directly compete alongside the system, undermining it with no net environmental benefit.

Amongst other sectors, the EED will force industrial and power sectors already covered by the EU ETS to reduce energy consumption by 1.5 per cent per annum. Power companies will be expected to help their customers to reduce their consumption by providing assistance to buy energy-saving devices such as double-glazed windows or insulation. This might appear positive on paper, but by forcing such a reduction, companies will be automatically emitting fewer gases and, consequently, will require fewer credits for meeting ETS targets. If this occurs, the price of carbon credits will reduce, meaning that other sectors, covered by the EU ETS but not by the EED, will be able to take advantage of the situation and buy extra credits if required at a lesser cost. Conversely, those making emission savings will see less of a return on their surplus credits. All in all, the overlapping regulations will mean incentives to invest in low-carbon technology will drop considerably. This is not to mention the rise it will cause for domestic energy bills which the EU has admitted will be a major source from which power companies will meet the additional costs.

This has not escaped the notice of EU technocrats and there has been grumbling from them: something extraordinary given the convention of not discussing draft legislation. Peter Vis, the chief of staff of the EU Climate Commissioner, suggested the EED would 'undermine' the carbon market. He said:

> More than half of those measures [in the draft] target the installations covered by the emissions trading scheme… We have got two policy approaches knocking up against each other and that isn't helpful… We're big supporters of energy efficiency but we have to be careful not to undermine a system that is in place now – the ETS.[15]

According to sources obtained by Reuters, the implementation of the EED will see prices during Phase III fail to rise and even fall slightly, hovering around the €14 mark for the 2013-20 period, nowhere near enough to have any impact on low-carbon investment. A second study suggested the outcome would be even more extreme, and that a crash would occur *à la* Phase I, with the price of permits falling to near €0.00.[16] The overlapping EED is not good for business or the environment and fails to address the real problems of the EU ETS: if the EED is implemented, the EU will be taking a step backwards on environmental policy.

While the aviation sector is almost unanimously opposed to compulsory inclusion in the ETS, it is in favour of a 'single European sky'

Within the wider context of reducing carbon emissions, improving energy and fuel efficiency is a more appealing idea than the EU ETS, and has more backing from industry. While the aviation sector is almost unanimously opposed to compulsory inclusion in the ETS, it is in favour of a 'single European sky'.[17] This proposed scheme would unify all existing independent regional and national airspace into a single European airspace under the control of the EU. Not only is it thought that this is safer and would allow enlarged capacity, but it would also increase flight efficiency. According to the UK Transport committee:

> Flights in Europe are on average 49 kilometres longer than needed. The European Commission estimates that the fragmentation of Europe's airspace costs €1 billion each year and that shorter, direct routes could save five million tonnes of CO_2 annually.[18]

This is an ideal form of policy: an economic and environmental win-win situation.

UK and EU funding solutions: investment funds

Given the inability of the EU ETS to deliver steady flows of cash for use in government-driven low-carbon investments, Britain (in the form of the Green Investment Bank [GIB]), and the EU (via the Marguerite Fund [MF]), are both aiming to bring public and private ventures together. On the face of it, this is a good idea, and helps to relieve some of the problems related to investment risk as discussed above, but they have been implemented poorly and suffer from the same short-termism as the ETS. Both the MF and the GIB are aimed at satisfying the immediate targets already shown to be ineffective.

The Marguerite Fund has currently raised €700 million, and expects to increase this to €1.5 billion by the end of 2011. Alongside promoting private sector network investment, the MF states its objective as:

> To combat climate change and contribute to the implementation of the EU's 2020-20-20 climate and energy targets which need to be met by 2020, in particular by supporting renewable energy technologies.[19]

It aims to invest between 25 and 35 per cent of its total budget in the energy sector and a further 35 to 45 per cent in renewables, with the rest spent on transport and networks.

Similarly, in the UK, the GIB will in theory receive £3 billion over the period 2011-15:

> The GIB's mission will be to accelerate private sector investment in the UK's transition to a green economy. It will play a vital role in addressing market failures which are holding back private sector investment… sectors likely to be eligible for intervention initially include offshore wind, non-domestic energy efficiency and waste.[20]

Having said that, the GIB is enduring a myriad of delays and general foot-shuffling. Both funds are suffering from entirely the wrong approach, and pigeonhole low-carbon investments without satisfying the more important investment requirements of long-term technologies. For the MF, this is perverse: its term is 20 years, so it should be looking to develop energy supplies that will mature in the same future timeframe, not existing measures that will be redundant by 2031. The GIB is evidently not free to make its own choices on where to invest, or is suffering from poor leadership: why else focus on high-cost-for-low-return offshore wind power?[21] Moreover, it will only be able to offer initial start-up capital, not the long and sustained investment required by most companies.

PART THREE

Prescription: Investment in Low-cost, Low-carbon Energy

CHAPTER 7

Low-cost, Low-carbon Energy as the Goal

- Low-cost, low-carbon (LCLC) energy allows the EU to reduce its emissions while remaining financially solvent: a win-win situation.
- The aim of policy should be to develop LCLC energy so that plentiful energy is available without an environmental cost.
- Current policy aims to drive up fossil fuel energy costs to incentivise low-carbon energy generation. LCLC energy development takes the opposite approach and aims to drive the cost of low-carbon energy down to fossil fuel levels with carbon taxes, in order to incentivise its adoption. Once this threshold is crossed, demand for low-carbon power will become self-sustaining and will overtake that of fossil-fuels.
- Energy supply accounted for 39 per cent of British CO_2 emissions in 2010, a figure roughly in line with EU output as well. By decarbonising the power sector, Britain would be half-way towards meeting the 2050 target.
- So far, the government has only spent £12 million per annum on investment in renewable energy R&D, the equivalent of 20p per UK citizen. However, state investment in LCLC energy is necessary to stimulate private-sector investment confidence. This could be direct, as state-aid rules allow low-carbon investment, or through a state-backed investment institution. The allocation of funds must be free from bias to allow it to support whatever the most promising LCLC technologies are.
- There is a huge international market that would purchase LCLC energy generators. Exporting this technology to the developing world is a highly effective way to reduce their emissions while they are increasing energy consumption in the process of development.
- Without global action on emissions, especially from China and India, all emission reductions made by the UK and Europe will be marginal in their impact. Exporting LCLC energy is the only way to make a real difference.

If the EU ETS is dismantled, there is the question of what, if anything, should take its place? Clearly, something is required, as the reduction in carbon dioxide emissions will not happen without some sort of government intervention: few companies would be willing to burden themselves voluntarily with extra costs. The failure of the ETS's market-based nature proves companies look for ways out of these costs if they can.

The answer to the above question is twofold. Firstly, existing targets are not seeking to maximise emission reductions at the minimal economic cost. The goal should be changed: to drive the cost of low-carbon energy down to truly competitive levels as fast as possible (although this is still years away). Secondly, the method to fund this intensive R&D should be a carbon tax, initially at least levied on the same installations covered by the EU ETS and for the same reason, that they are the biggest emitters.

The consumption of energy is not inherently bad for the environment. Nor indeed is a rising consumption of energy. It is only when the demand is met through a carbon-intensive supply (as via fossil fuels) that environmental damage occurs through the emissions. If the fuel is low-carbon, as renewables and nuclear power are, then as much energy can be generated as needed to oil the cogs of the economy, as cheaply as possible, guilt-free. The ultimate goal is therefore low-cost, low-carbon energy, henceforth LCLC energy. The size of the ambition here is clear: we should be aiming to drive the cost of LCLC energy down to the same price as existing fossil fuel energy sources *without* carbon costs. This leads us to some very clear objectives:

- To reduce the cost of low-carbon electricity generation to 6-6.5p per kWh.
- To reduce the cost of domestic low-carbon unit prices to 13p/kWh for electricity and 4p/kWh for space-heating and water-heating, the market led by gas at present.
- To reduce the cost of industrial low-carbon unit prices to 7.5p/kWh for electricity and 2.3p/kWh for heating.

Despite its huge reach, the EU ETS does not attempt to tackle this issue, but rather attempts to reduce the quantity of emissions produced, a goal that becomes far easier to achieve if cheap, low-carbon energy is available. The EU ETS is not enough to deliver this and will not incentivise emission-reducing investments either. We should set our sights higher.

The EU ETS is a mechanism that effectively circumnavigates the point: the main aim is reducing CO_2 emissions and the development of renewable energy sources is hoped to be a consequence of this, rather than a separate aim in its own right. Hence, instead of channelling money into the R&D of low-carbon innovation, it requires additional supports to achieve this. It would be far more efficient, in the short and long term, for money to be injected into research now, with the expectation that costs of production will then fall in the future, as has been the case with so many other technologies. This would ensure that fossil fuel technologies lose their current price advantage and would initiate the swap over that the EU ETS can only indirectly influence.

The reasoning behind low-cost, low-carbon energy

The Professor of Energy Policy at Oxford University, Dieter Helm, has asked: 'Is the way in which the monies are spent the most cost-effective way of achieving the desired outcome [of] low-carbon emissions?' and his answer is 'almost certainly "no"'.[1] We do not have unlimited resources with which to reduce our carbon dependency, and this means some doors are closed to us when considering what is the best solution. The EU cannot, for instance, price energy-intensive industries out of the economy and expect to remain financially buoyant. Nor can the EC allow EU ETS revenues to be creamed off by big business and carbon profiteers. If what is currently being done is relatively cost-inefficient, the alternatives must be examined. Of all of these, low-cost, low-carbon (LCLC) energy is by far and away the most effective solution. Successful economies are based on plentiful, cheap energy, and there is no reason why this strong foundation should not continue, provided it is low-carbon to boot.

At present, it is assumed that if a reduction in emissions is desired, then focusing on raising the cost of high-carbon energy is a means to this end, rather than decreasing the cost of low-carbon power. The increasing cost of energy experienced over the last few years is the result of naturally rising fossil-fuel prices, combined with various green charges like the EU ETS that artificially raise the cost further. The EAC has stated:

> Through interventions in the market and complimentary policy measures, using the full range of fiscal and policy instruments available, the Government should drive up the price of carbon steadily to a level where renewable and low-carbon investments become economically viable.[2]

What seems to have been forgotten here is that everyone wants (and feels entitled to) as cheap an energy supply as possible: all that needs be changed is that this supply should be low-carbon.

> ***...everyone wants...as cheap an energy supply as possible: all that needs be changed is that this supply should be low-carbon***

This is the most sensible approach to reduce emissions where they matter the most. In 2010, energy supply accounted for UK GHG emissions of 191.3 $MtCO_2e$, while industrial processes produced just 8.6 $MtCO_2e$.[3] In percentage terms, energy supply therefore accounted for 39 per cent of British CO_2 emissions, a figure roughly in line with EU output.[4] In other words, by decarbonising the power sector, Britain would be half-way to meeting the 2050 target. In EU terms, it is expected that by 2050 the European power sector, which produced 1.2 $GtCO_2e$ in 2010, will produce just 0.1 $GtCO_2e$ in 2050, a huge reduction of 95 per cent.[5] This cannot be achieved with anything less than a paradigm shift in how we generate electricity, and no level of energy efficiency based on fossil fuel can deliver this. The biggest emission reductions will therefore be made through decarbonising energy production, not through making energy consumers more efficient.

...by decarbonising the power sector, Britain would be half-way to meeting the 2050 target

Focusing on LCLC power is likely to propagate many more benefits than an equivalent investment in energy efficiency. For a start, once the power source has been changed, this requires no shift in the consumer's habits or usage: they can continue to consume power at the same, if not a higher rate, because emissions are irrelevant. Energy efficiency drives are often reliant on the opposite approach, needing consumers to alter their own patterns in a manner few are willing to accept in the short, and possibly the long term. Without their cooperation, nothing will happen and as the easiest habits are altered, such as not leaving lights on, others, such as persuading people to use their tumble drier at 3 a.m. become much harder. Politically, the fight also gets tougher, and the more alienated the electorate feels, the less enthusiasm will be felt for the environmental agenda overall. Most of these 'little changes' also have an imperceptible effect, even on a national scale. If all UK households switch off their phone chargers when not in use over a year, this would save just 0.25 per cent of their homes' electricity.[6] To get anywhere near real energy savings, the number of cumulative changes would have to be huge. Moreover, expectations that energy efficiency moves will drive down usage often ignore new or post-recession resurgent demand. Even DECC has plotted out the likelihood of future electricity demand, which currently stands at around 60GW peak, and found that by 2025 demand is likely to be between 55GW and 75GW.[7] This is not a favourable outlook for cutting back our usage, so the best option is to ensure that whatever we use, it is as low-carbon as economically feasible.

Technology may help in eliminating waste and improving efficiency, but this is limited. Electricity-wise, there is no easy paradigm-shift: even the most sophisticated nuclear reactor generates electricity by boiling water to drive a steam turbine. Physics has not yet delivered a better means, although the potential exists. For example, a thermocouple causes a current to flow between two zones of different temperature, but industrial-scale exploitation of this effect remains prohibitively expensive. There are other measures available for energy efficiency as well. Combined heat and power, by which waste heat from power stations may heat nearby homes, has long been popular in Scandinavia, and there is scope for reducing fuel use in transport by measures ranging from lighter vehicles to more efficient engines. Insulation and draught-proofing at home can also require less energy to heat a room. Although these savings can be genuine and permanent, they often come at an initial price that is too high for one of the parties. An individual must pay for and install a condensing boiler, saving on gas bills, but potentially paying for an earlier replacement. Driving a lighter car saves fuel at the risk of the driver's coming off worse in a collision. In power sector projects, the price of energy-efficiency, though it offers lower running costs, can mean that the actual selling price is pushed above that of competitors. A house built to high standards of insulation can be rendered more expensive, but the purchaser may not be swayed by the prospect of savings.

Despite the front-loaded costs, businesses are prepared to invest in worthwhile ventures. Where energy efficiency makes genuine savings on fuel costs, many businesses would be prepared to endorse it without needing a further subsidy. The incentive is to save on the cost of power, of which only a small proportion is the embedded cost of carbon from the ETS. A carbon tax is in line with this incentive. As far as energy-efficiency improvements are possible, practitioners can save the whole cost of fuel, tax and all. That incentive existed before any carbon-trading scheme came into operation. When improvements get harder, it remains possible to save the tax component of fuel costs by using a low-carbon energy source.

Crucially, economic growth must not be forgotten in the drive for low-carbon energy. A lot has been said already about the potential for 'green jobs' within the low-carbon economy, and for the most part, this has not been supported by evidence. At the current rate, green jobs will be created, but will be outnumbered by the losses incurred in other areas of the economy, such as energy-intensive industries. Even the EC's own research shows Britain will suffer net job losses of between 10,000 and 30,000 workers if current renewable-focused policies continue.[8] Switching to a LCLC power focus is clearly the way to maximise economic benefits on the back of environmentalism. If manufacturers have access to cheap, plentiful energy, the savings – compared to current green policy-inflated energy prices – will be significant, and can be passed on to consumers, greatly increasing

European manufacturing is not dead... Europe's highly skilled workforces are already a significant draw for many companies... If the installation of LCLC occurred, Europe would continue to dominate the industrial landscape for many more decades

their competitive advantage. In the current context, this is vital. European firms are up against BRIC rivals who already pay low prices for energy, sometimes even below the market rate; hence the increasing levels of imports and steady decline in exports. The only way to reverse these trends, and that of overall industrial emigration by companies wishing to take advantage of cheap energy, is to play them at their own game. European manufacturing is not dead, and many countries such as Germany will do everything they can to ensure their industries never weaken. Europe's highly skilled workforces are already a significant draw for many companies, for whom energy is not a major overhead. If the installation of LCLC occurred, Europe would continue to dominate the industrial landscape for many more decades.

The government's role

The greatest barrier to real low-carbon energy investment, let alone low-cost, low-carbon, is the lack of a long-term guarantee of commitment. This is nigh on impossible to provide as it relies on a policy continuity at odds with politicians, whose capricious nature and whims are risks that cannot be mitigated against. Regardless of the science, there is no way to know that environmental regulations will be just as important a decade from now, and while long-term green laws have been created, they can be repealed just like any other legislation. The solution is to provide hard evidence, in the form of significant direct funding, and to position low-carbon power as economically, not just politically, attractive.

With this goal in mind, it is the responsibility of the government to help bring it about: as the failures of the EU ETS show, the private sector cannot be relied upon to act as hoped for if left to its own devices. Volatile prices and short-termism have undermined existing low-carbon risk mitigation and, if we could dispose of the ETS, the provision of state funds injected into LCLC R&D, sourced from a carbon-tax, is the way to go. Far from being politically unlikely, the previous government considered subsidies to be acceptable alongside a rapidly tightening emissions cap, where the government must 'fill in the gap' between the carbon price and the actual cost of implementing low-carbon technology.[9] This would be highly inefficient.

The direct funding of LCLC power generation research should be indiscriminate, and be provided whether or not the source is renewable (thereby including nuclear energy). According to Dieter Helm:

> In the current European context, [economic] growth... could come through greater investment and from improving technology – and in the climate change context through investment in existing low-carbon technologies and through developing new technologies, through research, development and deployment.[10]

The ideal policy would be one that funded the rapid development of new energy sources and drove down their costs so that the whole economy and every household could benefit from the result. Access to cheap energy should be just as important for a government to provide as access to quality healthcare or a good education.

Public opinion clearly stands by this. A recent opinion poll of April 2011 revealed that 'energy security is the leading environmental issue for Britons, over and above climate change' and 50 per cent of Britons 'feel that future energy supplies and sources is one of the most important environmental issues facing the nation'.[11] Britain was the third most concerned of 24 countries about energy and 18th on the issue of climate change. As Figure 7:1 (see p.76) shows, this is unsurprising. The cost of domestic energy in the UK in 2010 was above the average EU price, despite the fact that it was less than the average as recently as 2006. This rapid price hike is primarily due to the extra charges added on by measures such as the EU ETS and the inefficient subsidy of renewable power thus far.

All existing low-carbon power generation policies have been reliant on pinning low-carbon subsidies to existing electricity prices which are driven by fossil-fuel costs. Taking wind power as an example, it has been claimed that:

> [W]ind generation costs have fallen dramatically as more wind generation has been deployed, to the point that they are already competitive (on-shore) with coal and gas generation, *providing the latter pay an EU carbon price of €30/tonne* [emphasis added].[12]

This is all very well, but it relies on the €30 cost. This has never been achieved in Phase II and will take years to appear in Phase III, if it does at all, so clearly wind power is *not* competitive. The ethos behind this is one of raising the relatively cheap cost of fossil fuel energy to levels comparable to high-cost, low-carbon ones. Figure 7:2 (see p.77), is taken from a report for DECC by Mott MacDonald and displays this idea, showing that even with an added carbon costs, gas is still cheaper than onshore wind and even the most carbon-intensive coal is cheaper than the least expensive offshore wind. Even by 2023, after over a decade of development, offshore wind is still projected to cost £110/MWh, almost £30/MWh more than the current cheapest fossil fuel.[13] It is clear that penalising fossil-fuels does not automatically mean the market will switch to and invest in low-carbon fuels, and it risks a continued high-carbon technological lock-in. The only way to ensure this will not happen is for the government to be the driving force behind the technological jump.

Figure 7:1
A comparison of electricity prices for household consumers in the UK and Europe

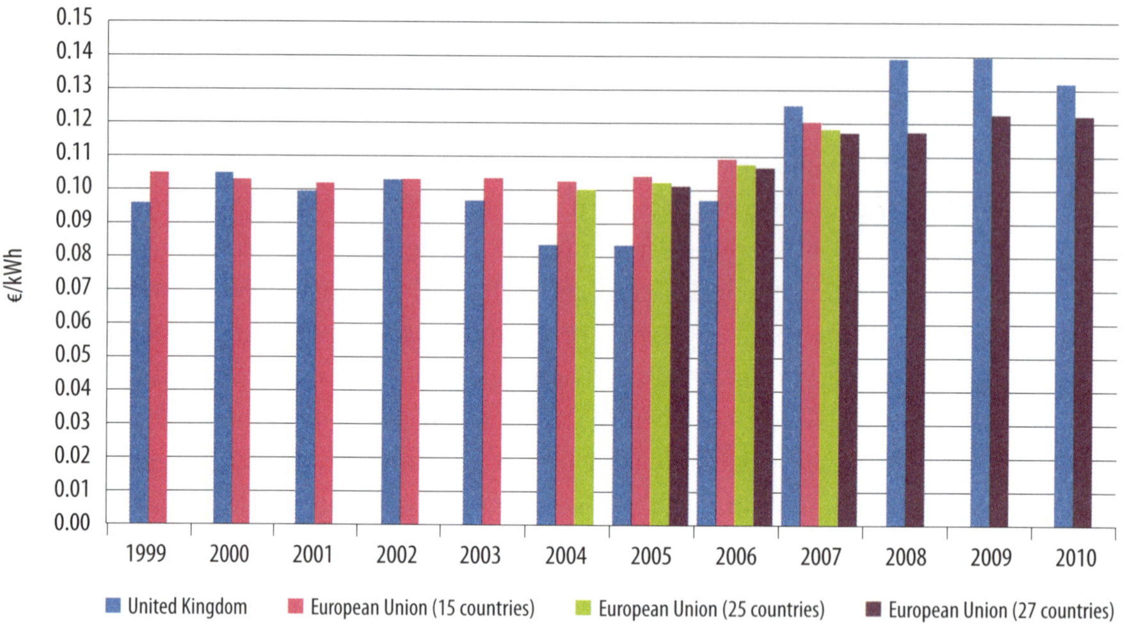

Source: Eurostat

Direct investment is the only way forward. Dieter Helm used the example of the trial carbon capture and storage (CCS) coal plants:

> If the demonstrations are left to the private sector, then the incentive is the patenting of the technology, but this limits the public good of knowledge diffusion. And since it may be hard to capture the technology for the specific company, the likely scale of technology development will be slow.[14]

If a company is unlikely to monopolise the resulting technology, its enthusiasm for developing it will be lower. This is unsurprising, but must be overcome by kick-starting the process. Without significant government assistance there will be little motivation to expand into new fields. For example, the Porto Tolle CCS project in Italy was designed

Figure 7:2
The cost of generating electricity with respect to carbon dioxide emission costs

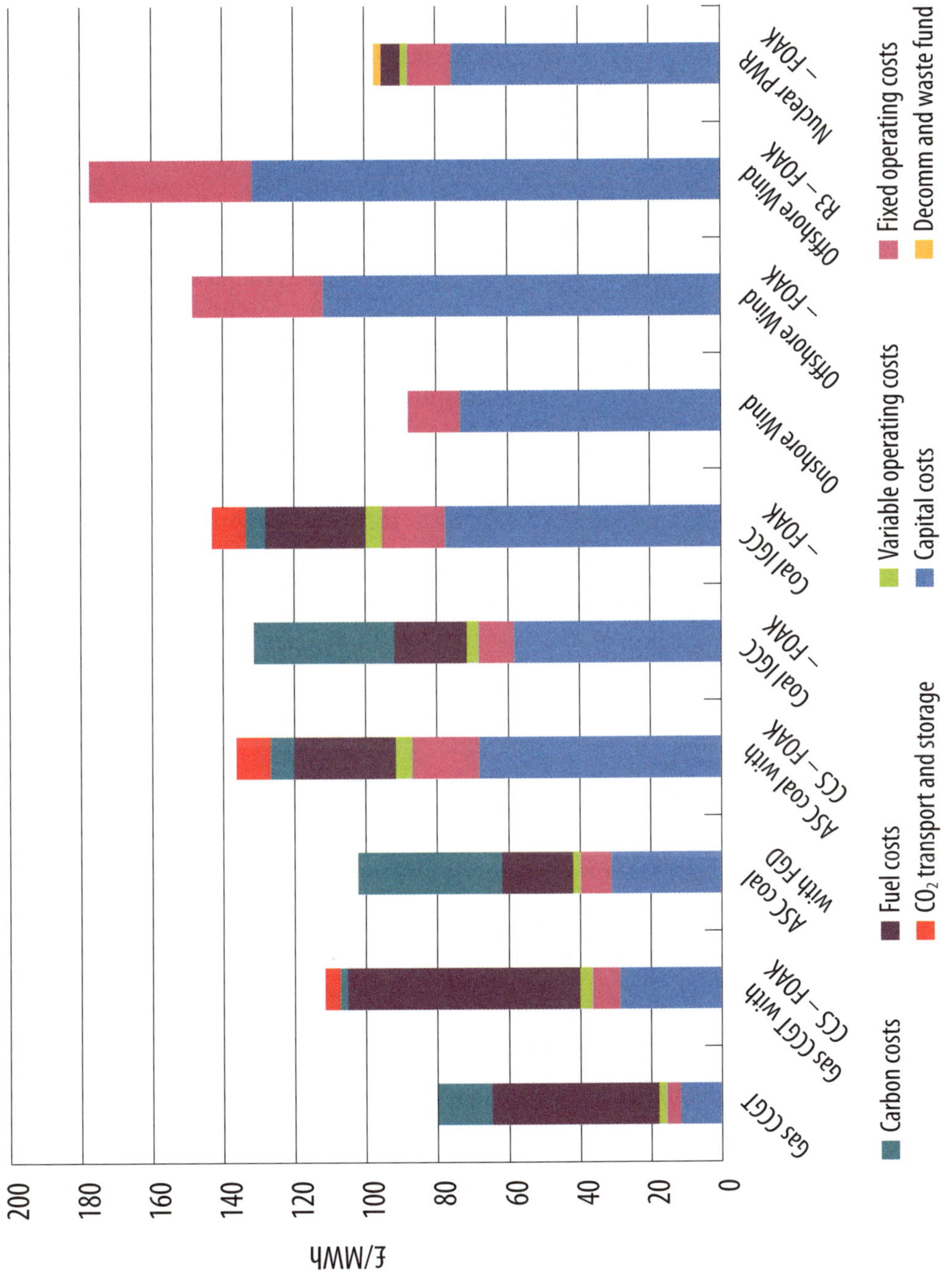

Source: Mott MacDonald, *UK Electricity Generation Costs Update*, June 2010, Table B.1, Case 2

to convert an oil-fuelled power plant into a coal-fuelled one, with an eighth of the output fitted for CCS. The project received €100 million from the EC but also secured €1.8 billion in private funding.[15] (In the end, the development was blocked by the Italian High Court, in part after environmental group pressure.) The case demonstrates that there *is* minor private enthusiasm for CCS (an eighth is a small investment) but this has to be balanced with the practicalities of also producing conventionally priced power to offset the higher costs involved. Current financial incentives are not enough and, in the CCS example, the EC is reserving 300 million EUAs to reward mature projects. This is all very well, but relies on funding for the initial construction and successful running being found.

The opposite approach to existing policies is needed, one which channels effort not into artificially raising carbon costs, but intensively and rapidly brings down the cost of low-carbon energy. So far, the government has only spent £12 million per annum on investment in renewable energy R&D, the equivalent of 20p per UK citizen. Instead, it prefers to pay out subsidies for whatever inferior renewables are available.[16] This gentle trickle of subsidies so far has unsurprisingly failed to have much impact and continuing these ineffective offerings will be a waste of taxpayers' money. Instead, the self-proclaimed 'greenest government ever' should directly inject much larger sums into low-carbon R&D, to speed up the process of making low-carbon power viable and desirable without artificial props.

Funding methods

Once the revenue for low-cost, low-carbon energy research has been raised through a carbon tax (see p.93), the next hurdle is to decide how this should be allocated. There are two principal methods, either indirectly through existing channels or directly by subsidising research. Which is chosen depends on whether the plan is to provide repayable loans or simply hand out subsidies to the research, with no intention of reclaiming the money in the future, no matter how successful the innovation is.

As discussed above (see p.64), in their present form the Green Investment Bank and Marguerite Fund are not suitable for delivering LCLC energy development, but in a post-EU ETS Europe, their roles could be re-examined. If the proceeds of the carbon tax were to be the source from which the GIB and MF are funded, this would ensure a steady flow of cash into them which could then be invested in any projects that showed the potential to deliver LCLC energy. Over time, as the returns of the tax decline (as in theory carbon emissions will be reduced), the initial investments will begin to show a return, generating a sustainable future source of further investment funds. This form of investment bank has already run very effectively in the UK, in the form of the Industrial and Commercial Finance Corporation, which was dedicated to improving the industrial capacity of post-war Britain. It gave funds to any firm it felt merited these, on a case-by-case basis, with decisions made predominantly by experts within industry, not bureaucrats or politicians.[17] It goes without saying that it would be crucial for any such modern institution to be independent of bias, and this includes independence from environmental pressure groups, so as to allow it to support the low-carbon technologies most likely to yield low-cost power, regardless of their political favour.

Alternatively, if it is perceived as more desirable for the EU and the government to invest directly, they can do so freely. Initially, it might appear that this would fall foul of EU state aid regulations, outlined in Article 107 of the Treaty on the Functioning of the European Union (TFEU). However, state aid rules do allow for state funds to be used to subsidise research, development and innovation (R&D&I) especially where such research has wider environmental benefits. In a way, this is perfect. It means that if the government were to finance R&D in nascent technologies, the products could be funded until the point at which they became commercially viable. There would be no need for companies to introduce them into the market to cover their costs while still being developed, as has occurred with wind turbines. Instead, the entire innovation process could be completed before commercialisation, at which point the government could reassess the need to continue to provide funding, either ending then or temporarily maintaining it, if it is felt to be necessary.

Export potential and the benefit for developing countries

> *Europe will become the beating heart of low-carbon evolution and can export the means to LCLC power globally.*

The goal of government environmental policy should be a very clear one: for the generation of low-carbon energy to become cheaper than the generation of power from the cheapest fossil fuel without carbon taxes. Once this threshold is crossed, demand for low-carbon power will become self-sustaining and will overtake that of fossil-fuels. As this happens, low-carbon investor confidence, so highly valued in the EU ETS, will increase and become entrenched in a far more efficient manner. This is good not just for the EU, but the whole world: Europe will become the beating heart of low-carbon evolution and can export the means to LCLC power globally. While extra-EU countries have been involved in this research as well, theirs has been based firmly within what is already possible. For example, China's solar panel and wind turbine industries are the largest in the world, dominating via their low costs. However, they are firmly rooted within present-day technology, and are of questionable quality and efficiency. Europe may not be able to compete on price with these rivals, but it certainly has the potential to dominate a future energy-production market based on high quality. If this can be mastered, then the initially higher capital expenses will become less of a concern. Given the existing specialisations of certain EU countries in certain LCLC energy sources, this would also mean that Europe would develop a wide portfolio of technologies to meet different requirements.

This potential export would also genuinely help the same countries the CDM is failing to aid. The CDM and other mechanisms attempt to circumnavigate the issue, by providing various forms of funding for projects but these only tackle the symptoms of industrialisation: factory or power station emissions and forms of pollution. It is not designed to deal with the much larger cost of shifting the foundations of developing economies away from cheap, carbon-intensive energy. While it might be possible for Europe to consume less energy, and consequently reduce its emissions, the same cannot be said for the developing world.

No country which is just beginning to bring its people out of poverty will be content to reduce their standard of living by raising the cost of energy

Current environmental thought still expects developing nations to bear the large economic cost of reducing their emissions, understandably something few are willing to do. No country which is just beginning to bring its people out of poverty will be content to reduce their standard of living by raising the cost of energy. Indeed, these nations will require increasing and cheaper levels of energy, to develop economically and improve lives further. For them, carbon emissions are understandably a secondary concern. The EAC theorised that in 2050, to meet a global reduction of 80 per cent of CO_2 levels, even if the OECD nations are producing near nil emissions, this will only allow others to emit 2-2.5 tonnes of CO_2 per capita.[18] This is a hugely ambitious target: the average Western country currently emits 20 tonnes of CO_2 per capita. For an assumed 89 per cent of the global population to live on a tenth of this is unlikely, especially since, as the failure of the 2009 Copenhagen climate change summit showed, a global agreement is unlikely any time soon.

By developing LCLC energy, Europe would be helping developing nations far more effectively than through the CDM, and indeed many other forms of existing international aid. There would be no need for tiresome and ineffective treaties or promises, and the power sources will sell themselves.

A global solution without empty promises

International environmental meetings and treaties have often fallen flat, as some countries refuse to do anything, while others sign up to targets that are either unrealistic or that they have no interest in keeping. The EAC has stressed that 'failure to reverse the rise in global emissions before 2020 could render much of the UK's domestic action meaningless'.[19] This failure looks likely to occur. China and India alone accounted for 9.8 billion tonnes of CO_2 emissions in 2009, a rise of nine per cent on 2008, regardless of their renewables investment: China has been doubling its wind power capacity every year for the last five years. In comparison, the EU generated 'just' 3.1 billion tonnes of CO_2 in the same year, so the cuts the EU ETS can offer are not significant on a global scale.[20]

Of course, there are many ways to reduce global emissions within a global context, but universal reduction is not something we should expect to happen. At present, despite the need for imminent action, recent international negotiations over carbon deals have floundered. This is partly because they have focused on developing countries promising to halt the growth in their emissions at some point in the future, which is obviously good, but not the reduction that is ideally required. More pressingly, no country wants to commit to reducing their emissions unilaterally, with the associated economic costs: only the UK appears prepared to do this and will suffer accordingly. In an ideal world, a global agreement would be made, and all countries would be imposing the same costs on their industries, which would negate any loss of profit or rise in prices, as everyone would have to deal with these. In reality, it is a leap of faith to hope that other countries will impose similar limitations on competitive advantage, and it will be problematic if they fail to do so. With no nation ready to jump first, deals grind to a halt, as has now happened with the EU's talks about moving to a 30 per cent reduction in emissions.

Focusing on developing low-cost, low-carbon energy offers the ideal solution to this quandary. BRIC nations may or may not take their emissions seriously and their commitment to reducing them relies on economic development not being undermined in the process, hence the ambiguous responses to global treaty propositions. If British or European companies are able to develop and export low-cost, low-carbon energy

If British or European companies are able to develop and export low-cost, low-carbon energy facilities, then there is no need for agreements or treaties and market forces can be relied upon to spread the developments

facilities, then there is no need for agreements or treaties and market forces can be relied upon to spread the developments. This will curb emissions much faster and much more effectively.

Energy security

The development of LCLC UK energy supplies is less of a luxury than a necessity. Britain will soon be facing an energy crisis that may jeopardise our already fragile energy security, courtesy of the EU's 2010 Industrial Emissions Directive (IED), the EU's successor to the 2001 Large Combustion Plant Directive (LCPD). The IED, which must be integrated into UK law by 2013, sets new targets for the levels of various harmful emissions that installations are allowed to emit. It covers 52,000 plants in total, many of which are already part of the EU ETS. In particular, it will have a major effect on the power sector, and will force plants either to comply with its emission targets to continue running; refuse and therefore close by 2016; opt out and close by 2023, running for a maximum of 17,500 hours until then; or, finally, to continue running under a national transition plan which will ease in the targets, although this relies on the country adopting such a plan and there is no guarantee the UK will. Given the UK's reliance on fossil fuel power stations, and coal in particular, this could have huge implications. At present, 15 per cent of power plants, specifically oil and coal ones, are due to close by 2016 anyway as a result of the original LCPD. However, due to the IED, further closures will occur and DECC has estimated its impact:

> One of the key elements of the IED is that it affects any gas plant commissioned before 2002, which is most of the UK CCGT fleet. This means that up to 40GW of existing coal and gas plant could be affected. That said, a number of plants are likely to retrofit abatement equipment to reduce their emissions and comply with the new Directive.[21]

...a quarter of our entire energy supply could vanish in just five years

In total, the IED could force an additional ten per cent of power stations to shut, meaning a quarter of our entire energy supply could vanish in just five years. Should this occur, the price of energy will soar even further, affecting industrial and domestic users alike. Whatever happens, it is clear that the days of many existing power stations are numbered.

Unless the government is content to let Britain import its electricity, there is a real need to develop new energy sources fast. In theory, the gap will be partially filled by existing renewable sources. As part of the EU's 2009 Renewable Energy Directive (RED), the UK is supposed to derive 15 per cent of its energy from these by 2020. In 2010, only 3.3 per cent of British energy consumption came from renewable sources, a rise of 0.3 per cent on 2009.[23] The overall cost of meeting the target is estimated at £100 billion by the *Financial Times*.[23] Ultimately, the choice is currently between impoverishing the nation by obeying the IED and RED, or recognising that existing power stations and new fossil fuel ones are necessary in the medium term to prevent the otherwise inevitable blackouts. Quite simply, these two targets are impossible to achieve with our current technology if we intend to keep the lights on. Given the unfeasibility of satisfying RED, the only available options are to adjust it, to a lower, more realistic expectation, or simply to ignore it and face any penalties this creates. It would be most sensible simply to scrap the RED: as already discussed, provided the energy is low-carbon, its renewable nature is irrelevant.

...the choice is currently between impoverishing the nation by obeying the IED and RED, or recognising that existing power stations and new fossil fuel ones are necessary in the medium term to prevent the otherwise inevitable blackouts

Developing LCLC energy sources is the long-term way to avoid both these problems, although not in the short timespan currently being worked to. Existing low-carbon technology cannot reliably deliver power on the scale we require and is too inefficient. To meet the RED target would require between 30,000 and 75,000 wind turbines to be constructed by 2020, a figure too high and too expensive and consequently too damaging to our economy.[24] It would be better for resources to be spent on accelerating LCLC development and for us to deal with a little altered *status quo* until then.

CHAPTER 8:
The Timespan of Real Decarbonisation

- The real aim of emission targets is not the 2020 target but the 2050 target. In the UK, we should be aiming to make the 80 per cent reduction by 2050 feasible instead of rushing the 34 per cent by 2020 target, which undermines our long-term ability to meet the later target.
- Existing targets promote the use of inefficient technologies currently in existence, no matter how expensive they are. This is the discredited 'picking winners' policy applied to energy.
- LCLC is the overall goal, but the government is not the mechanism for deciding how this should be reached, only the wallet from which the cash flows.
- A far more open-minded approach is needed, with a positive business and research environment allowing investment to flow into whatever technologies appear to have the greatest potential to deliver true LCLC energy around 2030.
- New generation nuclear power looks likely to offer LCLC energy and to have the potential to produce 100-300 times more energy than existing reactors, while consuming most of their own waste in the process.
- Thorium power must receive LCLC investment as its potential is so high. The basic cost of a ton of thorium is $300,000, but this quantity provides 1GW of energy per annum, sufficient to power a large city. For the same power, existing plants require 250 raw tons of uranium at an annual cost of $50-60 million. Thorium reserves will last 60-80,000 years.
- Given the UK's island nature, marine technologies such as tidal and wave power have great potential, but are attracting less investment than they should given the high start-up costs and the lack of state willingness to co-invest.
- In blue-sky thinking terms, concentrated solar power (CSP) offers a way to power the entirety of Europe using just four per cent of the Sahara Desert. Even in less ambitious terms, it is hoped CSP-generated electricity will halve in cost to $0.06/kWh by 2015, making it as cheap as gas and coal power are today, ignoring any carbon price.

The EU ETS is designed to deliver emission reductions *asap* and revolves around the 2020-20-20 concept. While well-intentioned, this misses the point of the actual aim of climate-change legislation: to reduce emissions significantly by 2050. In the UK, the target is an 80 per cent reduction in emissions on 1990 levels. It is this long-term goal that should be the motivating force of any policy, and all regulations should focus on this, not intermediate targets. The EAC noted:

> It is not vital to meet any specific budget in any specific year as long as cumulative emissions are limited on the way to meeting longer-term targets. But there is a risk that the UK could meet the near-term targets and budgets and still fail to deliver on the longer-term targets or to bear down on cumulative emissions…
>
> …It is important that the Government focuses action not only on meeting the carbon budget in any one year but also on taking action now to ensure that targets and carbon budgets can be met in the medium to long term.[1]

The EU ETS is at significant risk of letting the EU fall into this trap as it has distracted political discussion from the main environmental target. It is designed to stop businesses emitting CO_2 through efficiency measures, but is not able to provide the long-term investment required to see zero-carbon industry feasible.

Through their short-term nature, the 2020-20-20 European targets prevent the cheapest and most carbon-effective solutions being put into action. This slogan stands for a 20 per cent reduction in carbon emissions on 1990 levels, 20 per cent of electricity to be generated from renewables and a 20 per cent decrease in energy consumption through energy efficiency, all by 2020. These percentages and the year chosen are completely arbitrary, selected purely for their rhetorical potential, not because of any scientific premise. Hence, what they actually do is promote the use of technologies in existence today that are much less efficient than those currently in development, and which cannot be deployed in the period. Instead, the obligations require a 'quick fix' of whatever is currently available, no matter how expensive they are. This has led to the grossly inefficient reliance on offshore wind power, which according to Dieter Helm: 'is so expensive that it makes even nuclear power look cheap'.[2] The market would never have chosen this source were it not for the subsidies on offer, which tilt the playing field downhill. While developing better LCLC energy sources might pay off in the long run, the offer of quick returns has been acting as a dangerous distraction for many companies.

While not doing so explicitly, the EU has effectively been 'picking winners', translating the disastrous form of industrial policy into equally ruinous environmental terms via the Renewable Energy Directive. Rather than directing the form of technology that should be used, we should instead be ensuring the whole long-term R&D environment is fertile by moving away from funding based on short-term goals. LCLC is the overall goal, but the government is not the mechanism for deciding how this should be reached, only the wallet from which the cash flows. The correct policy is the same as that of a successful economy: create the conditions for vibrant and successful sectors. This would truly let the market take care of the problem, using the same principle as the EU ETS, but in a much more effective manner, where entrenched technological advantages are irrelevant and the potential for a paradigm shift in equipment is acknowledged: whatever has the greatest potential to deliver true LCLC energy in the long-term can deservedly attract the greatest investment.

Critics of the late-developer approach have suggested that leaving the problem for now will make tackling it in the future harder. The CCC has said:

> Equal percentage reductions from 2006 to 2050 would require CO2 emissions in 2020 to be 39 per cent below the 2007 level and 43 per cent below the 1990 level. This could be seen as an ideal benchmark: anything less in 2020 means that the challenge in subsequent years is increased.[3]

This statement is technically true and the easiest emission cuts are being made first, making later ones harder and more expensive. However, the argument fails to take into account the larger and faster low-carbon technological advances that will be made in the coming decades. It is highly likely that the ability to reduce emissions will accelerate, as it already has done for many years such as in the field of photovoltaic cells, which has come on leaps and bounds in the last decade alone. With sustained commitment and government investment, there would be no reason this should not continue. All in all, we should expect an initial rush of carbon emission reductions, as easy solutions such as scrubbers are implemented. This will be followed by a slower period, in which R&D is yet to catch up with society's desires, before the promised land of LCLC energy delivers the most significant emission reductions.

The emphasis on speed is a problem that pervades the government's policies. The fourth carbon budget was announced in May 2011, making the UK the first country to commit to long-term reductions post-2020. The carbon budgets, which began in 2008, are detailed in Table 8:1. According to this, within less than 16 years, the UK is expected to reduce its carbon emissions by 50 per cent, meaning that within the twenty years after that, by 2050, only a 30 per cent reduction is needed! This is the reverse of the situation we should be expecting, and in all likelihood these ambitious targets, like the RED legislation, will not be met.

Table 8:1 UK Carbon Budgets

	Budget 1 (2008-12)	Budget 2 (2013-17)	Budget 3 (2018-22)	Budget 4 (2023-2027)
Carbon budgets (Mt CO_2e)	3,018	2,782	2,544	1,950
Percentage reduction below 1990 levels	22	28	34	5

Source: DECC website

The ethos in environmental policy has always been to look not very far ahead. Presumably, in 2020, the next wave of legally binding targets will be set for around 2030. This is simply the wrong attitude to take as it prevents the support for green initiatives that are unlikely to deliver in the specified timeframe, hence the wind power fetish. It could be argued that this is irrelevant, that R&D projects such as CCS will eventually be rolled out in an economically competitive form regardless of state funding, but if this occurs a decade later than it could do, the opportunity cost for not intervening is significant, given that we will be left with mediocre technology in the meantime. According to John Constable of the Renewable Energy Foundation:

> [T]he cost to date [of meeting the Renewables Obligation], from 2002 to 2010, amounts to approximately £5.6 billion, with the oncost to 2020 adding a further £39 billion, a sharp increase in consumer burden. If we assume that after 2020 no further efforts are made to expand capacity, but that subsidies are maintained for capacity already installed under the RO, a further £60 billion would be added to bills, with the result that the total cost of the scheme from 2002 to 2030 would amount to £100 billion.[4]

£100 billion is a huge sum to pay for anything less than the entire upgrade of the British power supply. For instance, it is twice the cost of fitting CCS to all gas and coal power stations. It is also roughly double the capital value of all UK electricity generation, after depreciation, and is even close to the gross replacement-with- new value for all power station types, including nuclear and wind. This is quite a waste given the post-2020 LCLC technologies which are on offer but ignored. Below are just a few of the most promising technologies capable of delivering LCLC energy in the long-term, if sustained investment allows for development and deployment. They are listed in no particular order of preference.

Generation IV uranium fuelled nuclear power

Nuclear power is low-carbon, and even taking construction, maintenance and other carbon costs into account, it is still comparable to renewable power in terms of carbon produced per kWh. The next generation of nuclear reactors, Generation IV, are currently in development, with an estimated deployment date no earlier than 2030. These plants will be vastly more efficient than the UK's existing, mostly Generation II fleet. At present, the running cost of British nuclear plants is comparable only to fossil fuel plants with carbon costs added (10p/kWh) and most of this price (8p/kWh) is due to the capital costs associated with nuclear reactors. Decommissioning and nuclear waste management constitute 0.2p/kWh.[5] For a wider discussion of nuclear costs, see Appendix 6 (p.133). The existing forays into next-generation nuclear power, such as the Westinghouse Advanced Passive 1000, have already produced attractive results that seem likely to reduce capital costs significantly, through reducing the amount of engineering required, to hopefully bring the cost down to the holy grail of fossil fuel power minus carbon costs. Efficiency in current developments has also been increased. Fast breeder reactors have already been shown to produce a 60-fold rise in efficiency compared to standard once-through reactors.

> *If uranium could be extracted from seawater… the technology becomes highly sustainable and, using fast breeders, would allow the whole world of seven billion to produce 360kWh per person per day for 1,000 years, compared to 0.55kWh per person per day at present*

Future developments are even more attractive. If uranium could be extracted from seawater, where 99.41 per cent of global uranium reserves are found, the technology becomes highly sustainable and, using fast breeders, would allow the whole world of seven billion to produce 360kWh per person per day for 1,000 years, compared to 0.55kWh per person per day at present.[6] While this technology is not yet deployed, and has a cost roughly five to fifteen times more than extracting uranium from ore at present, the cost will fall significantly in the future if demand is sustained.

Other forms of still theoretical fourth generation plants look likely to be even more efficient, producing 100-300 times more energy, and consuming most of their own waste in the process. Indeed, they will even be able to consume existing nuclear waste, turning it into a valuable resource rather than a hazard.[7] In addition, the small amounts of short-lived waste they do produce will be proliferation-resistant. Clearly, they will address many of the concerns currently raised about the viability and safety of nuclear power.

Thorium power

The potential of thorium nuclear reactors is of particular note. The specifics have been documented elsewhere, but suffice to say, reactors will be powered by the element thorium, rather than uranium.[8] This has some significant advantages, alongside the low-carbon potential. Thorium is more abundant: global deposits of thorium would last approximately 60-80,000 years.[9] It is more common than tin, mercury or silver and is four times more abundant in the Earth's crust than uranium. Safety risks are marginal, with thorium reactors, often among those classed as Generation IV, designed to make meltdown impossible. In terms of efficiency, the basic cost of a ton of thorium is $300,000, but this quantity provides 1GW of energy for a year, sufficient to power a large city. For the same power, existing uranium plants require 250 raw tons at an annual cost of $50-60 million. Given the length of time that thorium can remain inside the reactor, this is highly promising. Their physical size is also tiny compared to existing reactors.

Safety risks are marginal, with thorium reactors… designed to make meltdown impossible

In terms of development potential, the UK's National Nuclear Laboratory delivered the following verdict:

> It is not envisaged that thorium fuel in light water reactors will be established in the next decade, but could be feasible in the following ten years if the market conditions are conducive.[10]

The long-term prospects are tantalising and, using an accelerator system, thorium could deliver 120kWh per day per person for over 60,000 years.[11] On this basis, thorium is a clear frontrunner to receive R&D funding for post-2020 power.

Tidal power

Tidal power is a renewable energy source, which has growing potential to deliver electricity at low-demand times of the day. Tides have been harnessed for their energy potential for millennia, but in terms of electricity generation, current barrages are not particularly efficient. The development of the new generation of efficient turbines is still ongoing, and the construction of the dams required can take up to ten years, so, to all intents and purposes, it should be classed as a post-2020 technology and treated accordingly. One interesting development is the tidal turbine, which is a freestanding underwater structure, very much like a wind turbine in aesthetic. However, this would be more efficient than wind turbines, require fewer materials and reduce any impact on the environment.

So far, there have been few UK ventures into tidal power, primarily because of its inability to deliver immediate returns. For example, a £3.5 billion scheme mooted for the Mersey Estuary was cancelled. According to the assessors of the project, Peel Energy, this was because:

> In the longer term, once the upfront capital costs have been paid off and for the rest of its 120 year life, the cost of electricity would be very competitive. But the preferred scheme is unlikely to attract the necessary investment while the emphasis in the financial sector and renewable energy incentives is on technologies that provide short- to medium-term returns.[12]

...the Severn Barrage, which was scrapped in October 2010, could have generated five per cent of the UK's entire energy needs

The high R&D costs mean that, without government help, this technology is unlikely to get off the ground. Similarly, the Severn Barrage, which was scrapped in October 2010, could have generated five per cent of the UK's entire energy needs. Concerns over its ecological impact were perhaps overshadowed by the £30 billion price tag, but concerns on both fronts could be overcome with further investment and development. It was for this reason that, the Energy Secretary 'did not rule out the possibility of the barrage as a longer-term option in the future, if market conditions improved'.[13]

Wave power

The harnessing of wave power out at sea is also under development, and along with tidal energy may mean that marine power can make a significant contribution to the UK energy mix post-2020. The UK is already at the forefront of wave technology, having the largest amount of wave energy companies in the world by some margin. The long-term development potential is therefore quite high. For example, AWS Ocean Energy of Inverness has proved its concept sufficiently to attract Alstom, the French engineering giant, to take a 40 per cent stake in their company.[14] One of the latest designs to catch attention is being developed by the Kent-based firm Checkmate SeaEnergy.[15] It is known as the Anaconda, but is more scientifically called a distensible tube wave energy converter. It will be a rubber tube up

The UK is already at the forefront of wave technology, having the largest amount of wave energy companies in the world by some margin

to 200m in length which is anchored just below the sea surface, as waves pass over it. The product prototype will be developed by 2014 pending investment, but full commercial production is still years off. Similarly, Pelamis Wave Power has been testing its own eponymous machine, which it hopes will see commercialisation by the end of the decade. At present, wave-based technologies are highly expensive and first generation wave farms will cost approximately 22p to 25p per kWh, a cost that will fall over time.

So far the development of marine renewables has to an extent relied on public money via initiatives such as the Marine Renewables Proving Fund, which provided £22.5 million to leading technologies. This is a small sum compared to the cost of actually bringing wave power to market.

Concentrated Solar Power

Concentrated Solar Power (CSP) uses a large system of mirrors or lenses to concentrate solar heat onto a small area in order to drive a steam turbine and generate electricity. For obvious reasons the UK is not climatically suited to this type of technology, however there is a real opportunity for CSP to make a significant contribution to the UK's electricity supply through importation.

The development and deployment of large scale CSP power generation in North Africa are still in the planning stages, but the potential is enormous. The Committee on Climate Change stated in their 2011 Renewable Energy Review that 'in theory CSP could meet all of Europe's electricity demand in 2050 using around four per cent of the Sahara desert'. The most advanced proposal so far has been DESERTEC which is a €400 billion, 40-year project which aims to provide 15 per cent of Europe's electricity needs by 2050.[16] This will use high-voltage DC submarine transmission lines to deliver the power to Europe, as this is much less lossy than conventional AC power transmission and a proven technology which already links France and England.

...in theory CSP could meet all of Europe's electricity demand in 2050 using around four per cent of the Sahara desert

Some estimate that by 2015 the cost of CSP-generated electricity will halve to $0.06 per kWh, making it as cheap as gas and coal power are today, ignoring any carbon price.[17] Therefore CSP looks to have significant potential to become an important LCLC technology.

All these above technologies are, to a greater or lesser extent, in use already and all of these are inefficient compared with the potential they all have. The government should look beyond its short-sighted targets, and no longer insist that inferior forms of these or other technologies are deployed, but invest in these specific types, knowing that, given time and freedom from political pressure, they can deliver the LCLC energy we need.

CHAPTER 9

The Means to the End: a Trans-EU Carbon Tax

- The carbon tax would translate into economic terms the perceived harmfulness of CO_2, with all revenue being directed towards long-term emission reductions via LCLC energy R&D. The tax itself is a stick, promoting less carbon-intensive fuel use, but the end result is a carrot, as the investment will deliver reductions in low-carbon energy costs.
- The tax should replace all other European and national green levies businesses are paying, so that it is the single environmental charge they pay. Imposing the carbon tax at the EU level is the perfect way to ensure uniformity of costs and therefore harmony of impact, avoiding carbon leakage and economic damage. However, if the EU refuses to abandon the failed ETS, Britain should do so unilaterally and adopt the carbon tax as the sole green tax.
- The tax promotes the development of the most carbon-saving technology available, while the ETS favours the implementation of the bare minimum investment possible.
- The tax provides a steady price, delivering the investor confidence the ETS by its nature can never provide and does not allow scope for lobbying, favours or corruption.
- It is a more long-term solution, allowing the price to be adjusted if new scientific research reveals that CO_2 emissions are more or less pressing than previously thought.
- The current aim of the Government's carbon price floor is to charge £30/tCO_2 by 2020. If the flat-rate carbon tax were set at £30 per tonne of CO_2 emitted, this would generate revenues of £7 billion from EU ETS installations and £2 billion from domestic gas users on a yearly basis for investment in LCLC energy research. More detailed research into the economics of such a tax might suggest that a lower level would be preferable.
- All additional fuel poverty caused by the tax, and all fuel poverty caused by previous green policies, could be negated by investing most of the first year's revenues in the Warm Front scheme.

The goal of this report is to show a path to a future of large-scale low-cost energy that is also low in CO_2 emissions. Before that goal is reached, there will be some unavoidable economic pain, but that pain can be minimised far more effectively than at present and needs to be distributed as fairly and tolerably as possible. As stated, technologies for low-carbon energy exist already, but without the requisite R&D they will be neither large-scale nor low-cost enough to replace fossil fuels. Two issues stand clearly in the way of this. Firstly, we cannot predict what the optimum LCLC technologies will be a few years from now, let alone in a couple of decades. Our imperfect knowledge is not reflected in the EU ETS. Secondly, to speed up the development, much more funding is needed than the paltry sums generated from the EU ETS, which are not likely to be spent on green projects anyway. This hypothecation challenge would be overcome by the carbon tax because by its very nature it is a much more explicit stream of revenue aimed at very specific targets. To try to spend it on anything else would risk it being labelled a stealth tax and provoke resentment towards the government.

This report proposes that the research be funded from a tax on carbon emissions that would replace the existing green environmental costs at national and EU levels. This would translate into economic terms the perceived harmfulness of CO_2, with all revenue being directed towards long-term emission reductions via LCLC energy R&D. This would be intended first as a negative incentive, a stick, to use less carbon-intensive fuel. As the research delivers improvements, costs will fall, so that the incentive to use low-carbon fuels becomes positive, a carrot, when the unit costs are naturally lower.

> *The famous remark that fuel will one day be so cheap that it needs no metering remains a distant dream. Research to improve the balance of costs and side effects is a much more attainable goal.*

Energy costs extend beyond the price of the fuel. Wind and sunlight may be free, but then, in a sense, so are coal and uranium. The costs lie elsewhere, in the harnessing, extraction, refining, conversion, storage and distribution and in the handling of waste products. There are also social costs, such as the need to site plants near people. Different means of energy production have different mixes of costs additional to the cost of fuel. Fossil fuels have the head start in that their waste products have long been released to the atmosphere without apparent cost, whereas nuclear alternatives have quite properly been subject to cautious safeguards. The research needs to be directed to bringing down the costs wherever they lie, so that large-scale production is eventually cheaper for the fuels whose side effects are more socially acceptable. The famous remark that fuel will one day be so cheap that it needs no metering remains a distant dream. Research to improve the balance of costs and side effects is a much more attainable goal.

The advantages of a carbon tax over existing measures

In a perfect world, where information concerning the value and costs of LCLC power development was certain, there would be no need to argue the merits of a carbon tax over the emissions trading system as each would arrive at the same outcome. When the original carbon tax concept was rejected by the EU, this was primarily due to concerns about its subjective nature: there is no right cost and the level of emissions reductions cannot be known beforehand. Implementing it would have been a trial-and-error affair, whereas the ETS provided foreknowledge of reduction levels. As is now known, these estimated EU reductions do not translate into net global reductions and are not quite as clear as they appeared. Indeed, given the present inability to reduce emissions at the moment it would appear the price of credits is at an incorrect level that fails to incentivise any investment. At the very least, a carbon tax could not put the EU in a worse position than already exists.

Imposing the carbon tax at the EU level is the perfect way to ensure uniformity of costs and therefore harmony of impact. Aside from the obvious competitive issues raised by unilateral approaches, having a patchwork of different taxes would cause the overall scheme to be undermined, given that some, no doubt including the UK, would impose a higher tax rate on emitters and many countries would maintain exclusions that would be inconsistently applied across the EU. In addition, having uniform carbon tax rates means carbon leakage would be minimised as companies would have to relocate entirely outside the EU to avoid costs, rather than just move to neighbouring states as at present.

In foreign investment terms, this is also a key benefit: at present, the UK is losing out on extra-EU investments to other member states and the rest of the world as a result of its unilaterally higher environmental costs.[1] As a result, the tax should replace all other European and national green levies businesses are paying as well, so it is the single environmental charge they pay. This would ease the burden of regulation many companies feel is stifling, and reduce the overall green costs, whilst remaining at a level that encourages emission reductions.

Additionally, if the carbon tax is trans-EU, it would be much easier to create a border tax adjustment to ensure European businesses do not lose their competitive edge over extra-EU rivals who do not have to pay the carbon tax. Currently, there are no EU import tariffs to take account of the fact that firms in EU states have higher manufacturing costs due to their environmental levies, something companies in countries such as China do not have to bear. This means the EU is currently shooting itself in the foot, and making it easier for these rivals to dominate the manufacturing market. By imposing a border tax adjustment, this problem is negated, as the cost of these other goods is raised to take account of emissions likely to be produced in their manufacture. Given how far-reaching this would be, affecting many sectors, the trade-wars otherwise seen in specific sectors such as aviation would be avoided. The tariffs could easily and rapidly be reduced or cancelled if other countries imposed comparable carbon taxes.

The tax would be very likely to succeed where the EU ETS has failed. Unlike the ETS, a carbon tax actively favours investment in low-cost, low-carbon energy, which is the means to the greatest reductions in carbon emissions possible. The cap-and-trade approach clearly sets an acceptable limit on the level of reduction required in CO_2. This means that any innovation that decreases emissions faster than the cap is falling is likely to be undervalued, since the excess reductions will be seen as superfluous by the carbon market, which would not want to bear any greater cost than is absolutely necessary. The carbon tax takes the opposite approach and, by simply charging a set price, leaves the field open for innovations generating the most abatement possible, taking technological uncertainty into account.

In terms of the effect of the tax on businesses, the lack of price fluctuations is perhaps its greatest asset. This gives businesses the certainty they need to invest in green technology as they can be sure of how much they will save: a clearer incentive than is currently the case. Of course, the cost of a carbon tax is likely to be passed on to consumers, but this happens already with the EU ETS. It is an unfortunate but permanent side-effect of commercial carbon levies of any kind. What can be hoped for is that the cost to consumers remains stable rather than constantly rising, as at present.

The tax is also a longer-term solution to emissions: the EU ETS has a shelf life of only nine more years and only takes account of differing fossil fuels' carbon intensity by proxy of their emissions. A carbon tax could charge according to the differing carbon contents of various fuel sources. With no credits to trade in a tax, there would also be no need for the carbon market and no space for profiteers, delivering a refreshingly simple system. There would be no vested interests so lobbying could be kept to a minimum, and the focus of the system finally would return to the original aim of delivering emission reductions at the lowest economic cost. Professor Michael Grubb, chief economist of the Carbon Trust, pointed out:

> The most fundamental reason [why the EU chose an ETS over a carbon tax] is the one of the political economy… we have made progress with emissions trading where we could not with the carbon tax because it gives you an additional degree of freedom, namely how much re-allocation you have to give away to buy off the powerful lobbies.[2]

The *Economist* has also argued that getting the volume of permits wrong would have a much greater negative effect than misjudging the level of a carbon tax.[3] The latter can respond to new information or inefficiencies much better and without causing shockwaves and confidence collapses in the same way as altering credits or their price would. Being arbitrarily set, as further scientific discoveries reveal the speed and extent of anthropogenic global warming, this can be translated into economic terms by raising or lowering the carbon tax. Had a carbon tax been implemented in 2005 instead of the EU ETS, the pre-recession, recession and post-recession periods would have had their emissions regulated much more efficiently and according to their existing production levels. The glut of credits and current low incentive to invest in low-carbon technology would never have occurred.

The cost of running a carbon tax would be much smaller, for the EU and companies alike. Currently, time and money are wasted by firms having to ensure compliance and playing the carbon trading game. The former has already cost British companies £68 million in Phase I and has been estimated to cost £100 million by the end of Phase II.[4] On the other side, the pressing need for greater regulation of the carbon markets themselves is a huge cost waiting to happen, to add to the existing expenses of running the ETS. The tax bypasses all of this, and could easily be tagged on to the existing tax structures, meaning there would be little additional cost incurred in running it. This is also a much simpler approach. The only people who would lose out from this approach would be the banks, lawyers and consultants who rely on the complexity of the EU ETS to make their living.

The effects of the carbon tax

Most importantly, a carbon tax would provide a clear, stable, long-term source of revenue for low-carbon investment. With grandfathering the norm in the EU ETS, the Scheme is not a real source of income for government to plough into low-carbon innovation and, as already discussed, EU recommendations that governmental revenues are spent on green projects often go ignored anyway. A carbon tax would force governments to get their act together. Given that it automatically highlights what is being taxed and why, this also underlines what the revenue should be spent on. This transparency should make it more politically acceptable and moves away from the current negative view of green charges which are often seen as 'stealth taxes'.

The amount of money the tax could raise is quite eye-opening, even when applied only to the UK. The specifics of the carbon tax and how the following quantities were arrived at are all discussed in Appendix 1 and 2 of this report. The Government currently plans to charge £30 per tonne of CO_2 in 2020 via the carbon price floor. Suffice it to say, if the carbon tax were set at £30 per tonne of CO_2 emitted and no other environmental levies or taxes were implemented, this would generate revenues of £7 billion from EU ETS installations and £2 billion from domestic gas users on a yearly basis. This is roughly £1-2 billion more than the UK spent on low-carbon technology in 2009.[5] Whether the UK would be willing to pay £30 per tonne of CO_2, or would want a lower cost, is a discussion for elsewhere.

In terms of the costs to be paid from this revenue, operations charges will be minimal because of the tax working along existing channels already created. The major cost will be offsetting any rise in fuel poverty caused by the tax, and also that caused by the previous regime of environmental costs. This is likely to cost in the region of £7.5 billion, but the social benefits will be huge and, once dealt with, there will be little annual fuel poverty cost. Additional costs would include the ending of existing climate change policies. Of these, only the Renewables Obligation and Feed-In Tariff schemes involve long-term commitments which, whatever the desirability of the schemes themselves, should be honoured since individuals have made investment decisions of the strength of the assurances. The costs of this will be around £700 million per annum.

In total, the money available to LCLC research is therefore the sum raised from the carbon tax, less liabilities under the Renewables Obligation and Feed-In Tariff, less, for the first years, sums allocated to large-scale domestic improvements in energy efficiency. If the idea is to eliminate fuel poverty in a single year, this means there will be 'only' £1 billion available. If the cost of abating fuel poverty is spread, billions could be raised every year from the tax to be invested, creating a huge pot with which to ensure Britain becomes the centre of global energy research. The sums strongly suggest the fast-tracking of LCLC development is more than just a pipe-dream.

A target of cheaper large-scale energy for all may appear to leave little incentive for energy efficiency. When energy is cheaper, economic laws suggest that people will gladly use more of it. That is to forget that society has not yet reached that ideal position and needs to chart a course to get there. Proposing a carbon tax to fund research means that short-term energy costs will be higher although not by as much.

The carbon tax proposed here is intended, like every other tax, to be temporary, but unlike most others, carries a genuine prospect of its own extinction by providing a cheaper alternative to carbon-based fuels. While in force, it will cause unwelcome increases in prices, affecting industrial and domestic users. That is where energy efficiency becomes relevant.

Fuel poverty rises will be a consequence of implementing a carbon tax, although given the elimination of other green costs, these increases will be less than at present. However, unlike the EU ETS, the carbon tax could also alleviate this rise, and pre-existing fuel poverty. For the fuel poor, diverting a part of the tax's proceeds towards improving their energy efficiency is a direct way of alleviating the suffering. The beneficiaries have a particular interest in co-operating, since the improvements with the greatest effect are in heating, insulation and draught-proofing. A less draughty home is immediately more comfortable and takes less time to heat up. It is possible that after home improvements, residents will simply turn up their thermostats and spend the same amount on energy as before. At least they will have the choice. They will not be forced to shiver in the quest for low-cost, low-carbon energy, but if they choose to use more energy they will contribute extra through the carbon tax to the funds for research.

For industrial users, the market already provides incentives towards energy efficiency. Increasing the price, by whatever means, increases that incentive, but there is every reason to suppose that companies are already taking advantage of what savings they can. Energy represents a proportion of costs and ordinary market discipline dictates that reducing any cost brings a competitive advantage. Energy-efficiency improvements are far from futile, but it may be expecting too much to suppose that that they will deliver the hoped-for reductions in carbon-based energy use by themselves.

CHAPTER 10

Two Interim Solutions of 'Low(ish)-carbon, Low(ish)-cost' Energy

- We cannot wait two decades for the development of LCLC energy before upgrading our existing, ageing power sector.
- By 2016, up to a quarter of the UK's existing domestic power supply will be lost as a result of pre-determined closures and EU legislation. By 2023, all but one of the UK's nuclear power stations will also close, opening up another energy gap. Rapid deployment of existing technology with reasonable costs is desperately needed if we are to keep the lights on.
- Building new nuclear power stations would be the ideal mid-term solution, as it is effectively very low-carbon. France generates 78 per cent of its electricity from nuclear sources and, in 2008, France emitted 83g of CO_2 per kWh of electricity and heat generated, compared to 487g of CO_2 for the UK and an OECD average of 433g.
- The Government plans to build 16GW worth of nuclear plants to replace those closing, but the equivalent of a further ten Sizewell B plants will be needed to offset the more imminent closure of fossil fuel power stations.
- In addition, gas offers another attractive mid-term electricity source. If the UK were to switch entirely from coal to gas to generate electricity, this would save 34 $MtCO_2e$, which is a significant 5.8 per cent reduction of UK emissions at 1990 levels.
- Increasing gas usage could save the UK €100 billion between now and 2030, by which time fully-fledged LCLC could be brought on-stream.

As discussed earlier (see p.82), the combined effect of the Large Combustion Plant and Industrial Emissions Directives will reduce the UK's domestic power supply by up to 25 per cent from 2016, a situation that has not yet generated the political discussion it deserves. It is simply impossible for renewables, as they currently stand, to fill this gap: if we cannot meet a target of 15 per cent, where will the extra 10 per cent on top of this come from? It is time to be realistic, and recognise that something must be done to keep the lights on in the time before LCLC power is a viable energy source. Given that the delay before this is available is likely to be two decades if not longer, it would be wise to use as cheap (to the consumer) and low-emission fuel source as possible. Two solutions present themselves as ideal mid-term candidates: nuclear power as it currently exists and switching from coal to gas. Of course, economic and environmental goals are subject to political considerations, and this means the latter option would be easier to implement. In addition, there would also still be a need to maintain a mix of energy sources, and while nuclear or gas power would be emphasised, they would not be relied upon totally. Clearly, these two options are not mutually exclusive, and perhaps the best outcome Britain can hope for is the promotion of both gas and nuclear technologies through to viable LCLC energy joining the mix.

> *...the combined effect of the Large Combustion Plant and Industrial Emissions Directives will reduce the UK's domestic power supply by up to 25 per cent from 2016...something must be done to keep the lights on*

Less likely, more environmentally beneficial: increase nuclear power

Nuclear power *is* low-carbon, no matter what the critics say, and along with its low cost per kWh, as seen in Figure 7:2 (see p.77), this means it should be taken seriously as a solution to our imminent power shortage. As DECC itself puts it: 'nuclear generation currently reduces national carbon emissions by between 7 per cent and 14 per cent'.[1]

France is well known for its existing reliance on nuclear power, generating some 78 per cent of its total electricity from it, some of which is exported to the UK. This has had a huge impact on total power sector carbon emissions. In 2008, France emitted 83g of CO_2 per kWh of electricity and heat generated, compared to 487g of CO_2 for the UK and an OECD average of 433g.[2] The carbon cost of building nuclear reactors is often cited by critics as being very carbon-intensive, but this is simply untrue. David MacKay, of Cambridge University and now Chief Scientific Advisor to DECC, calculated:

> The steel and concrete in a 1GW nuclear power station have a carbon footprint of roughly 300,000 tCO2. Spreading this 'huge' number over a 25-year reactor life we can express this contribution to the carbon intensity in the standard units [of 1.4g/kWh(e)]... The IPCC estimates that the total carbon intensity of nuclear power (including construction, fuel processing and decommissioning) is less than 40gCO2/kWh(e).[3]

Nuclear power is also superior to developing increased gas power because it is more sustainable. Gas is used most efficiently to heat homes, not generate electricity. Given that there is a limit to the amount of gas that can be deployed at any given moment, a rise in its power generation use will leave less to heat homes. This hypothetical full capacity has not yet been reached, but it is something to bear in mind. At current usage worldwide, known mineable sources of uranium will last for 1,000 years, and even if the rate of consumption increases rapidly, these resources will never be exhausted before LCLC power is fully deployed.

> ***By 2023, all nuclear plants bar Sizewell B will have been closed. Given that in 2009 nuclear power generated 18 per cent of UK electricity, the loss of all but 1.4 per cent of this is another huge energy gap likely to open up.***

The decline of nuclear energy in the UK is a pressing concern. By 2023, all nuclear plants bar Sizewell B will have been closed. Given that in 2009 nuclear power generated 18 per cent of UK electricity, the loss of all but 1.4 per cent of this is another huge energy gap likely to open up.[4] In October 2010, the government announced plans to build a new nuclear power fleet of up to 16GW, funded by power companies without public subsidies. In the wake of the Fukushima nuclear crisis in Japan and the accompanying hysteria this was reviewed. At the time of writing, the government's response to the interim report by the UK's Chief Nuclear Inspector suggests that there will be little alteration to this policy.[5] If this new capacity comes to fruition, this will offset the pending closures.

So far, so good, but given the LCPD and IED will be reducing fossil fuel power stations by up to a quarter, this will require the equivalent of ten Sizewell B nuclear plants to make up for this loss, in additional to the 16GW planned, in as rapid a timeframe as possible.[6] The speed is not the barrier, and at its peak France was building 3.4 new nuclear reactors a year, so the IED's power debt could be filled providing the programme got fully underway.[7] The problem is more political, and despite the benefits of low-carbon reasonable-cost energy that nuclear power offers, the government does not seem to want to push for much more than replacing the plants the UK is already losing, choosing to keep nuclear-derived electricity supply constant. In part this might be due to the public perception of nuclear power, which is unreasonably fearful, a fear which is repeatedly preyed upon by the media. Professor Wade Allison of Oxford University has made the point very clear:

> New dangers are now evident. These are more global and threatening than any local nuclear incident, and arise from changes in the Earth's atmosphere, triggered by the continuing use of fossil fuels. Although many initiatives are possible in response, the only large-scale solution is a major switch to nuclear power for electricity generation and the supply of additional fresh water. For this to happen rapidly, cheaply and without disruption, the public perception of ionising radiation needs to be turned around.[8]

Perhaps along with the construction of nuclear power stations, the government could provide a public good through properly educating the public on the minimal risks nuclear energy poses to society.

However, a greater reliance on nuclear power would allow more fossil fuels to be removed from the energy mix prior to LCLC power maturing, an option that those concerned by carbon emissions should jump at in theory (and notwithstanding the other GHGs produced by burning fossil fuels). In economic terms, this is also the ideal option. David MacKay constructed one of his five 2050 energy mix plans around this, entitled 'Plan E' ('E' standing for economics, or the economically optimal solution).[9] With a 'strong carbon price' that could take the form of a tax, nuclear power would trump CCS and most forms of renewable to produce 110GW of power, the equivalent of France's current fleet doubled, with various renewables propping this up.

More likely, less environmentally beneficial: switch from coal to gas

With nuclear power's place in the energy mix not yet decided, but ageing coal plants already lined up to be decommissioned, there will soon be a looming energy shortage on the horizon. Something must fill this and, with the advanced LCLC power sources still years away from deployment, gas is the best candidate in economic and environmental terms to plug the generation gap and act as a bridge between fossil fuels and sustainable energy sources. The easiest way to balance these is to switch from the most carbon-emitting forms of fossil fuels to the least, or in other words, shift our energy source from coal to gas, which produces 50 per cent fewer emissions than modern coal plants and 60-70 per cent less than old ones.[10] For instance, combined cycle gas turbines achieve much higher thermal efficiency in power stations by harnessing the explosive energy of the gas and then recovering the heat produced as well. Size-wise, the chimneys can be built at half the height of coal station ones; cooling towers are also shorter and do not produce any vapour on all but the coldest days. In relative terms, gas is low(er)-carbon and, by increasing its presence in the energy mix, it provides us with a means to continue to work within the current technological paradigm without overly burdensome energy prices as well as emission reductions.

While it must be remembered that gas is a fossil fuel, and relying on it will sustain our fossil fuel dependency, this is acceptable, provided it remains just a medium-term solution and given the benefits it would bring. Within the European context, by relying on gas power generation as the main fuel within a mix of various sources including renewables until 2030, this could save €450-550 billion, depending on the future price of gas, according to a recent report by McKinsey, with the UK standing to save €100 billion from this.[11] If this is translated into domestic savings, this means a reduction of €150-€200 on household energy bills compared with those otherwise inflated by green costs. This also shores up the European energy-intensive industries, that would otherwise face a five to ten per cent erosion of their profits, risking 20-25 million jobs.[12] By holding off on renewables until 2030, their later implementation will reap comparatively greater emission reductions and it will still be possible to meet the 2050 80 per cent emission reduction target, having saved a huge cost in the meantime.

> *If the UK were to switch entirely from coal to gas to generate electricity... overnight, it would mean the UK would instantly meet and exceed the EU's 20 per cent CO_2 emission reduction target...That an option that can be implemented immediately, with no extra construction or investment costs, but instead has been entirely overlooked, is a travesty*

Switching to gas has already begun to reap environmental rewards and the UK has seen a 2.1 per cent decrease in GHGs produced per unit of electricity on this basis alone, as coal is abandoned in favour of gas.[13] There is the potential to take this much further. The load factor of existing gas plants rises when needed, primarily to compensate for the fluctuations in renewables' ability to generate power, but even then there is spare capacity. Were the plants used all year round, their output would roughly double and the grid would remain stable, and this extra power would allow other, older and more polluting coal plants to be taken offline. If the UK were to switch entirely from coal to gas to generate electricity, this would save 34 $MtCO_2e$, which is a significant 5.8 per cent reduction of UK emissions at 1990 levels (see Appendix 7 for details). If this could be done overnight, it would mean the UK would instantly meet and exceed the EU's 20 per cent CO_2 emission reduction target.[14] On an EU scale, this would reduce emissions by

250-300 MtCO_2e per annum, or 20-25 per cent of all total existing power sector emissions.[15] That an option that can be implemented immediately, with no extra construction or investment costs, but instead has been entirely overlooked, is a travesty.

This route is not one that forgets that the ultimate goal in emission reductions lies outside fossil fuel power generation. The gas switch merely provides Europe with the longer timeframe necessary to allow the development and deployment of these other technologies, to make LCLC energy as LCLC as possible. Using gas gives breathing space, at least until 2025, by which time it will be clear which options are most likely to work, rather than throwing ourselves on the wind turbines available because they are the only working renewable source we currently have. As advanced intermittent renewables come online in the future, gas can then shift its role to that of a backup source. Gas power stations can start up and alter their output quickly, so could also be used to shore up less constant power suppliers, something that will be needed for the foreseeable future.

In economic terms, the coal/gas shift makes sense as well. Compared to 2005, the annual average real-terms price of natural gas in 2010 has increased by 26 per cent; coal on the other hand has increased by 58 per cent. Coal prices are likely to continue to increase as demand from Asia grows. Gas prices are also likely to fall with the expansion of shale gas in the UK, where reserves could be as large as 150 billion cubic metres and estimates have already emerged predicting that over a tenth of Britain's gas needs could be produced from shale fracking by 2015.[16]

The key price for low-carbon energy to beat, so that it can qualify as low-cost as well, is the cost of generating electricity from gas. From DECC statistics, the prices paid for gas by power stations were as shown in Table 10:1.

Table 10:1 The cost of gas to power stations

Year (distinctive years only)	Gas price (p/kWh)
2000	0.591
2004	0.761
2005	1.015
2008	1.644
2010	1.461

Source: DECC, Average prices of fuels purchased by UK power producers and of gas at UK delivery points (QEP 3.2.1)

Converting to electricity at around 30 per cent efficiency means that the price per kWh of gas-generated electricity will be over three times as much. That element of the cost, before any climate-related charges, offers the potential for low-carbon energy to compete on price. When that starts to happen, the market will bring about the whole panoply of low-carbon benefits without the need for levies or taxes.

There are two key arguments against moving towards more gas usage in the UK: that its prices are volatile and we are reliant on other countries being able and willing to export it to Britain. Price volatility has been an issue in the past, with gas prices having spiked in 2005-07, shifting the balance back in favour of coal power. Ironically, due to these fluctuations, gas also suffers the same problem as EU ETS credits, namely low and unsustained investment. This means there is a real role for the government in promoting the shift to gas. Ofgem has made public this concern:

> Uncertainty relating to the impact of environmental policy makes forecasting future gas demand much more challenging for potential investors than might have been the case historically. This may delay investment in gas infrastructure that might be required should environment measures not fully deliver.[17]

In other words, should the overbearing focus on fast-tracking the Renewable Energy Directive not deliver, the UK has very little to fall back on and the impact of the chronic lack of investment has already been seen: the biggest threat to a gas shift is not the well-publicised dispute between Russia and Ukraine, which when at a peak in 2009 had far less of an impact on UK gas prices than the Rough gas storage facility fire in 2006. The gas price jump can be attributed not to the destruction of the facility itself, but to the fact there were barely any alternative storage and import structures. The UK has just 4.4 billion cubic metres of gas storage capacity, which is less than five per cent of UK annual demand, compared to other European countries such as Germany, which has enough gas storage to account for 20 per cent.[18] This is in part due to shifting market conditions. Previously, gas storage was viable as companies could buy gas when it was cheap in the summer, and sell it for more as prices rose in the winter. However this difference declined over time due to the increasing ability to ship in liquefied natural gas in super tankers during the winter. The government needs to step in and provide the incentive to develop storage, given that the risk of supply vulnerability will only get worse if gas use increases. Given the freedom from the burden of the EU ETS (if it is dismantled), it would be reasonable to oblige the big six power companies to construct capacity.

Energy independence, the critics of gas say, is of paramount importance and we should be able to produce power without leaving ourselves vulnerable. This is of course entirely true. However, there are three main reasons not to fear. Firstly, with the development of the means to extract shale gas, it looks as if Britain could soon be able to produce enough, onshore and then offshore, to power the nation without importing the fuel. On a European-wide basis, there are enough shale gas deposits to fuel the EU for 30 years at current usage, plenty of time to give other low-carbon energy sources the R&D timeframe they require. Secondly, being dependent on gas sources from outside the UK is not necessarily an energy security issue. Dr Jim Watson, the Director of the Sussex Energy Group at the University of Sussex, has argued:

> The popular belief that foreign sources of energy (including gas) are inherently less secure than domestic sources of energy is wrong – and is not supported by the evidence. Many of the security threats to UK energy supplies in recent years have come from within the UK. Examples include the Rough gas storage facility fire a few years ago and the truckers' blockade of oil depots in 2001.[19]

Thirdly, on a practical level, energy independence is an irrelevant point when weighing up the alternatives. As discussed above (see p.101), Britain is set to lose potentially a quarter of its energy generation by 2016, so if the enthusiasm for purely green power sources is maintained, Britain will find itself having to import so much additional energy from the Continent that it will be even more energy impotent. One way or another, our existing fossil fuel power stations have to be replaced by new but cleaner ones and gas is the closest thing to LCLC energy available for now.

A compromise

This report has identified two separate issues:

- The inefficiencies of the EU ETS compared to a carbon tax
- The need for low-cost low-carbon energy

While these could be dealt with independently, it would maximise efficiency if both points were looked at together. Nonetheless, it is possible to deal with the flawed current system, but only in the short-term, provided the second LCLC energy aim is adopted. Crucially, if the British government did decide to hypothecate the £4-8 billion in revenue of the EU ETS in Phase III, while this sum is smaller than the carbon tax's £9 billion, this hypothecation would raise the value of the EU ETS, on two conditions:

- The money should be spent on exactly the same things as the imagined carbon tax, i.e. LCLC energy R&D and combating fuel poverty.

- Other green charges would have to be scrapped: too many businesses would not be able to survive full exposure to them all. Given the comparable levels of revenue from the EU ETS alone, they are not all needed anyway.

This plan is a reasonable compromise that partially resolves the tensions investigated in this report. While the EU ETS does have its inefficiencies compared to the carbon tax, they could be overlooked temporarily if it did not damage businesses and it was known that the money was being invested for the long-term benefit of the country. While not the same as our recommended carbon tax, this would at least give something in the present system to work with.

Conclusion

Back in 2002, when the European Commission voted in favour of the EU ETS, its inception was heralded with strong words by its instigator, then Commissioner Margot Wallström:

> This is a landmark decision for the EU's strategy to fight climate change. It proves that the EU is taking action on climate change and gets emissions down, and that we do so in a way that minimises the cost to the economy. The world's eyes have been upon us to see whether we will succeed in creating the biggest emissions trading scheme world-wide so far. We have succeeded. It will help all member states, as well as the EU as a whole, to reach their Kyoto targets while cutting costs at the same time.
>
> As early movers on emissions trading we will gain valuable experience with this new instrument, and improve it over time.[1]

Needless to say, these expectations have proved overly optimistic and nine years down the line emissions have barely reduced, costs have risen and few improvements have been made. The ETS has never provided a concrete carbon price, and has instead piggy-backed on other green levies and the recession, to create a faint illusion of emission reductions. To paraphrase the Commissioner, we *have not* succeeded.

The root of the Scheme's problems is that it is not designed to be a long-term solution and is built on an unsustainable idea: it relies on some installations permanently polluting more than they should and therefore needing to buy allowances. If every installation were genuinely to curb their emissions below the capped level (the theoretically ideal outcome) then there would be no demand for carbon credits and the price would collapse. Such a price collapse would eliminate the economic incentive to reduce emissions further. By its very nature, the EU ETS is not sustainable.

The EC does not seem to have recognised the effect this has for industry. While the shape of Phase III up to 2020 is well documented, after this there is little concrete information about the future of the scheme. On the whole, the specifics have been organised at the last minute while companies are left second-guessing, with little time to react and adapt once decisions are made: the current fiasco with the inclusion of aviation is a classic example. The ignorance surrounding the post-2020 situation is highly disconcerting, especially since this is now only a few years away and fails to give businesses the time they need to invest accordingly. It appears that either the politicians don't know how far businesses plan ahead, don't care or don't even know what future environmental policy the EU will have. Even if the scheme is supposed to run indefinitely, the low costs of credit-buying combined with the lack of information means that firms are investing for the short-term and locking into carbon-intensive manufacturing for another business cycle. Without the clear political commitment to emission reductions, there is little reason investors will assume they have to adapt fully.

The EU ETS is not loved, even by some of the most ardent environmental groups who see it as a barrier to green investment rather than a catalyst. Such a view is entirely reasonable. The Scheme, by attempting to be all-inclusive, has crowded out alternative methods to reduce carbon emissions that would probably be more effective and efficient. Its wandering tentacles mean that the only way these alternatives can be implemented without economic ruin is by removing the ETS. Indeed, the Hartwell Group has stated: 'It is now plain that it is not possible to have a "climate policy" that has emissions reductions as the all-encompassing goal.'[2]

On paper, cap-and-trade schemes like the EU ETS do appear to be the best method to reduce emissions, but the number of vested interests, such as emission validators, carbon traders and industries lobbying for free credits, mean the system cannot work along market principles, except with a shadowy underworld of emission and credit falsification. The only solution to this is a tight level of monitoring, scrutiny, and carbon price floors, all seen as disadvantages avoided by using cap-and-trade in the first place. If this is required, then the scheme shifts to appear more of a carbon tax.

Indeed, mechanisms attempting to reduce GHG emissions suffer from their own version of Heisenberg's Uncertainty Principle. If the cap level is known, then the price cannot be. If the price is known, then the emissions output cannot be predicted. This is an inevitable problem, but given the emphasis is on the private sector investing in low-carbon technology to reduce their emissions, the priority has to be the creation of the right environment to ensure this will happen. The only way to do this is to create a steady price that can be predicted from years and business cycles in advance, and the only way to do this is a flat carbon tax. Not only would this create the crucial incentive to invest in low-carbon technology, which the EU ETS has not provided, but the revenue would be a clear fund which the government could use to subsidise green R&D or reduce business costs elsewhere, reducing the overall cost of abatement and shifting the image of a carbon tax from a penalty to an encouragement. *This* is the way to reduce emissions most economically.

The EU ETS may be legally binding, but it is also a failure and getting in the way of more carbon initiatives. The UK Government should realise this, retract its support and press for its dismantling: the Coalition's promise to be the 'greenest government ever' will be left unfulfilled unless it realises the ETS is not the most efficient vehicle to reduce emissions at the lowest cost. The best solution would be for the ETS to be scrapped and replaced with a carbon tax. Charging the same amount across the EU would ensure that no country suffers from carbon leakage and investors can be confident of similar treatment wherever they invest, overcoming the difficulty of carbon registries using different approaches. This is a sure way for the EU to know that it will be genuinely reducing its own emissions and does not give the scope for companies to try to offload their emissions elsewhere. To ensure UK firms' competitive advantage is not undermined, the tax should replace all existing green levies, including the carbon price floor, so it is the sole environmental charge businesses have to pay. This way, the government could be sure that it would not be pricing industries out of the UK, which is currently the case.

> *The EU ETS may be legally binding, but it is also a failure and getting in the way of more carbon initiatives*

The carbon tax revenues should not be the EU's or the government's to do with what they like, but should be exclusively invested in green projects aimed at delivering low-cost, low-carbon power, with a certain amount reserved to alleviate any aggravation in fuel poverty the tax causes. However, this level would be far less than currently likely under the ETS if the tax rate were maintained at a steady level that rose infrequently, as this would prevent power companies from claiming poverty and raising domestic bills: it should be the polluter who pays, not the consumer.

Since it was theorised, nuclear fusion has been dreamt of as the Holy Grail of energy sources, providing near-unlimited (and therefore very-low-cost), low-carbon energy once fully developed. However, it is clear this will not be deployed at any time soon, perhaps not this side of the twenty-second century. We must make do with whatever we currently have, but this does not mean we have to sell ourselves short: low-carbon power will develop to the point at which it becomes cost-comparable with fossil fuels even without pricing in the latter's carbon costs, at which point the market will embrace it entirely. In the meantime, the government should invest the carbon tax revenues in LCLC research to speed this process up, and provide a reasonable-cost mid-term solution. This also changes the ethos of the EU's approach to carbon emissions from one of accepting pollution providing penance is done, to positively encouraging its reduction.

...if Britain wants to provide the paradigm shift in energy generation the world of the future needs, no more time can be wasted: the EU's Emissions Trading System must go

There is no reason why we should settle for second-best; for corruption; for a costly scheme that fails to deliver on its promises. The environmental consequence will be that, the longer the EU ETS is allowed to continue, the lower our chances of reaching an 80 per cent CO_2 reduction in 2050. The economic consequence will be that, while we adhere to the ETS in a world in which no one else imposes stringent emission regulations, European industry dies a death that is neither heroic nor necessary. Together with other European nations, or alone, if Britain wants to provide the paradigm shift in energy generation the world of the future needs, no more time can be wasted: the EU's Emissions Trading System must go.

APPENDIX 1

The Carbon Tax – Options for Implementation

Upstream or downstream?

The tax could be applied in either of these two ways, but it would be more sensible to impose a downstream carbon tax. An upstream tax is applied to the carbon content of fossil fuels at an early stage of the supply chain, such as extraction or distribution. This means that almost all of the carbon entering the economy is effectively taxed and the price signal can follow through down the supply chain providing incentives to reduce CO_2 for all direct and indirect consumers of fossil fuels.

The primary advantage of this approach lies in its simplicity. It minimizes the number of points within the economy where the tax needs to be levied, simplifying the administrative process of ensuring tax compliance. In the case of the US, it has been estimated that all American CO_2 emissions could be covered by applying the tax to only 2,000 entities, in contrast to the downstream EU ETS approach applying to 12,000 installations, and under which not all CO_2 emissions are covered.[1]

However, this advantage is fatally undermined by the need to have a potentially complex system of exemptions and rebates. For the tax to be fair, partial or total rebates would have to be offered to those entities that can prove that some or all of the CO_2 contained within the fuels they purchase will not be released into the atmosphere. This would ensure that installations such as power plants with CCS technology or many industrial companies would not be unfairly penalised. In addition, CO_2 created as a by-product of chemical reactions, such as during the cement or ammonia production process, would not be accounted for and a separate charge for these would have to be applied, adding further complication, or the emissions would have to be ignored, making the tax unfair.

In contrast to the upstream solution, a downstream system would apply the tax to the actual emitters of CO_2, in a similar vein to the EU ETS. Indeed, targeting the same installations as the ETS is the ideal method, given the relatively small number of non-mobile emitters that can be monitored. The administrative cost will most likely be higher than under an upstream implementation, but the infrastructure for verifying emissions and ensuring compliance already exists as a result of the EU ETS. Using this pre-existing regulatory structure would help significantly reduce the initial costs of implementation. Therefore a downstream implementation seems the most effective and will be assumed for all calculations.

By targeting only a select few installations, this solution misses out certain other sources of emissions, such as road transport, but pre-existing taxes apply here and an additional charge on already burdened road users may be unpalatable.

Why do the statistics focus on Britain, not the wider EU?

Whilst the carbon tax should be implemented EU-wide, the calculations made here focus solely on the effect it would have for the UK. This is for two main reasons. Firstly, for the sake of simplicity: many EU states, like Britain, have their own environmental legislation on top of the EU Directives, and estimating the effect of removing all of these would be too prone to error. By using the example of the UK and removing its unilateral green policies, the results are clear and can be stated with some confidence.

Secondly, this report is at its heart concerned with the growth of the British economy and prioritises its health and emission reductions above that of wider Europe. This means that regardless of what the overall EU does, Britain should consider rejecting the EU ETS and unilateral costs and implement a carbon tax in their stead. The advantages, both environmental and economic, will become clear.

Who is included?

Initially at least, the tax should include the same installations covered by the EU ETS at present (but excluding sectors soon-to-be included such as aviation which, as discussed, will be too problematic to include). This is for the sake of simplicity and to ease the tax in, which could run via the same existing channels of revenue collection, leading to minimum disruption for the affected companies. The cost for power companies will be significant and these will trickle downstream to domestic consumers of electricity (although bill increases will be less than under the current regime of green policies). This means there is no need directly to impose the tax on the domestic electricity market, which would risk a 'double-whammy'. In the longer-term, once the tax has demonstrated its worth, it could be extended to include other, smaller companies and their CO_2 emissions, should this be felt necessary. The template for this already exists via the Climate Change Levy.

In addition, domestic gas supplies could also be included. These should not be included in the tax immediately, partially because the majority of CO_2 emissions come from the ETS installations, and also to allow businesses to adjust their business plans and households to regulate their gas consumption.

Given the vast revenues generated from the narrower initial scheme, the inclusion of other sectors would be unnecessary to fund LCLC energy R&D, but would primarily incentivise emission reductions, which might be politically desirable. Economically, given that the costs for almost all businesses would be less than at present, the effect on them would not be positive, but at least it would not be as negative.

What level should the carbon tax be set at?

The tax could be charged at any price per tonne of CO_2 emitted, but the necessity of maintaining economic growth and general levels of prosperity limit this cost. In the Government's investigation into the carbon price floor, three levels – of £20, £30 and £40 – were mooted. The middle-way of the £30 price floor was eventually chosen. It should be remembered that this was an additional cost on top of all the other environmental charges businesses are already paying.

We are not recommending the £30/tCO_2e level of tax but, given that the Government has been working with that figure, it has been used here as an illustration of future costs, although it would have to be implemented as the sole cost payable by installations. Even working with this figure, which was criticised by many industries during the CPF consultation period as being too high, it is clear from the evidence in Appendix 2 that the tax is significantly cheaper for businesses and consumers than current policies will dictate. This is not to say that a level of £30/tCO_2e will not have some negative consequences. Further investigation into the effects of such a cost will be needed, together with the consideration of lower carbon tax rates.

APPENDIX 2

The Carbon Tax – Statistics

Revenue

Estimates for the revenue raised by a £30 a tonne carbon tax assume that tax will fall on existing EU ETS installations and on domestic gas use. Given this, the estimate for revenue generated is around £9 billion, as shown in Table A2:1. It is acknowledged that this assumes a 100 per cent pay-up rate and no avoidance. Another assumption is that the tax is paid in full. Since the liability is calculated using the existing registrations for EU ETS and gas meters, this appears optimistic but not unreasonable.

Table A2:1 Revenue from the Carbon Tax at £30/tCO_2e

Emissions Source	**Million tonnes of CO_2**	**Revenue generated (£bn)**
UK ETS installations emissions	237	7.1
UK domestic gas	64	1.9
Total	301	9.0

The effect of the £30 carbon tax on domestic electricity prices

The tax represents a cost to electricity producers and it is reasonable to expect all or most of this cost to be passed onto domestic consumers in the form of higher electricity bills. In estimating this effect it was assumed that:

- For every MWh of electricity produced, 0.54 tonnes of CO_2 is released.
- Existing climate change charges and initiatives are adding around £55 to the average domestic electricity bill. This a conservative estimate, especially given Ofgem admits that the costs of some of these charges will rise significantly year on year.[2]

Given these assumptions, Table A2:2 shows what the percentage change on current domestic electricity prices would be if all existing green charges were scrapped and replaced with a carbon tax, varying by the level of the tax and the amount of the extra cost power companies pass through. Pass-through rates have been estimated at 'between 60 and 100 percent' for Germany and the Netherlands.[3]

Table A2:2 Present cost (£/MWh) to households of the carbon tax as a percentage change on current costs

Pass-through rate	Carbon tax rates			
	£10	£20	£30	£40
1	-6.7	-2.3	2.0	6.3
0.9	-7.1	-3.2	0.7	4.6
0.8	-7.5	-4.1	-0.6	2.8
0.7	-8.0	-4.9	-1.9	1.1
0.6	-8.4	-5.8	-3.2	-0.6

Table A2:2 shows that at a carbon tax of £30 and a high pass-through rate a carbon tax would leave domestic electricity prices more or less where they are now. Figure A2:1 (see p.118) shows the long-term effect of the carbon tax. At £30, while marginally more expensive than current policies when first implemented, this negative effect soon disappears (notwithstanding the tax revenues being used to offset any fuel poverty caused). By 2015, while the base price of electricity will rise, the cost of the carbon tax will not, so the comparative cost is now less than those of current policies, which have a greater impact.

In theory, the impact of a carbon tax would diminish over time as power companies respond to the incentive and reduce the carbon-content of electricity. However, for simplicity the effect of the tax on prices has been kept constant over time. Given the nature of power firms, it is also likely that bills to consumers, while they could not be continuously raised because of the tax, do not fall in line with savings either. By 2020 though, despite no reduction in the weight of the tax, the benefit over existing policies is even more evident. It is important to remember that these existing costs are a conservative calculation, given that there is no inclusion of any future green levies the government or EU might introduce into the existing mix. For domestic users, it is therefore clear the carbon tax is a winner.

Figure A2:1
Estimated impact of environmental policies and a £30 carbon tax on average domestic electricity prices

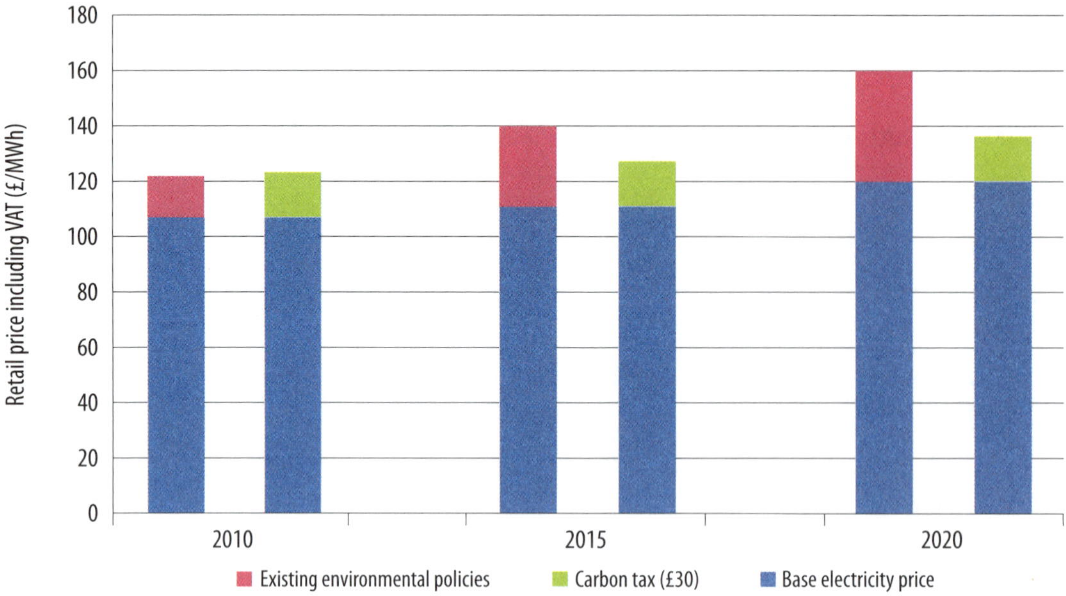

Source: DECC, *Estimated impacts of energy and climate change policies on energy prices and bills*, July 2010

The effect on domestic gas prices

If a £30 carbon tax were to be applied to natural gas, then this could increase consumer gas prices by over eleven per cent.[4] However by 2020 the DECC estimates that current environmental policies could add up to 18 per cent.[5] As seen in Figure A2:2 (see p.119), the results of this are similar if not as marked as the equivalent electricity tax.

As gas represents a larger amount of the energy consumed by households, the increase in gas prices outstrips any small fall in electricity prices we might see as a result of the introduction of a carbon tax and the removal of existing charges. This means that the overall consumer bill (assuming we are taxing gas and electricity at a 100 per cent pass through rate) will rise by around seven per cent if the tax regime were changed overnight, although in time this rise will become smaller than the rises likely under existing policies.

Figure A2:2
Estimated impact of environmental policies and a £30 carbon tax on average domestic gas prices

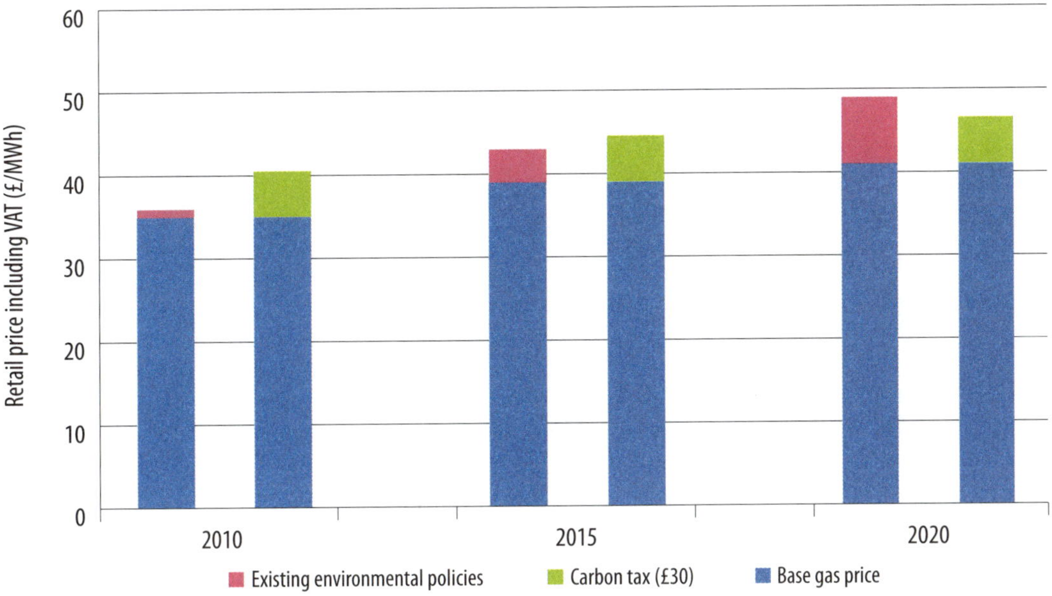

Source: DECC, *Estimated impacts of energy and climate change policies on energy prices and bills*, July 2010

The impact on energy-intensive users

The economic effect of high energy costs on energy-intensive users is discussed at length in the Civitas report *Chain Reactions*, so just the impact of the tax will be examined here. Given these industries are the most vulnerable to rises in energy costs, they are therefore important to consider when assessing the likely impact of a carbon tax on the competitiveness of the UK economy. Assuming constant consumption and carbon intensity of energy, Figure A2:3 (see p.120) displays the relative effects of current environmental policies, net of the estimated decrease in the carbon intensity of power generation caused by the Renewables Obligation and other regulations, and the effect of a £30 carbon tax with no assumption about a reduction in carbon intensity of power generation, on the price of electricity for energy intensive users.[6] Even with these unequal assumptions, we see that current policies will quickly cause the price of electricity for these industries to rise higher than when compared to a £30 Carbon Tax. The invisible benefit of the tax, its certainty, is also revealed here.

Figure A2:3
Estimated impact of net environmental policies and a £30 carbon tax on average retail electricity prices faced by large energy intensive users

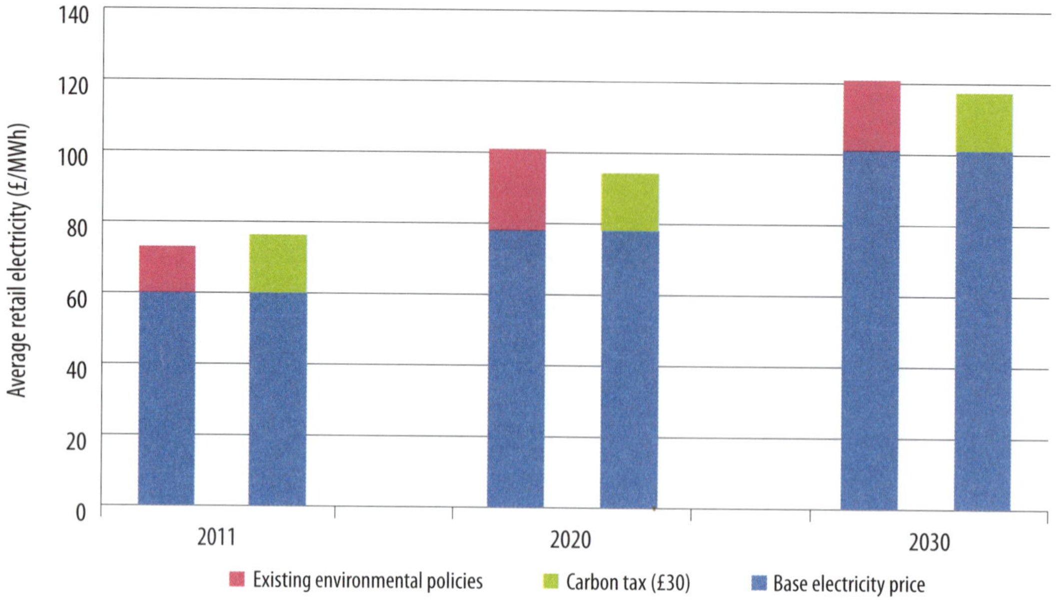

Source: DECC, *Provisional estimates of the impacts of energy and climate change policies on energy prices and bills of large energy intensive users*, Chart A1, Gas Price Unchanged scenario, July 2011

Effect on fuel poverty

Fuel poverty is defined as households spending ten per cent or more of their income on energy. As the above estimates show, the introduction of a carbon tax with any real impact could make domestic consumers *temporarily* worse off by having to pay more for their energy needs. A rise in energy costs has clear implications for fuel poverty. Unlike existing policies, which clearly exacerbate the fuel poverty problem, the carbon tax should be used as a revenue source with which to confront the issue directly, mitigating any potential increase in the number of fuel poor and reducing those already suffering from it.

Since 2000, £20 billion has been spent on fuel poverty schemes.[7] However, this money has done little to stem the rising numbers of fuel poor, and as of 2009 some 5.5 million households are still categorised as such, with numbers predicted to grow further as a result of green policies.[8] Existing government programmes have targeted the issue indirectly, such as via low incomes, through the winter fuel allowance, and addressed rising energy bills through engaging utilities companies to provide vulnerable households with social tariffs. However, the most effective and long-term policy measure has been the reduction in household consumption of energy through efficiency improvements, delivered through the Warm Front programme.

Warm Front

Warm Front, introduced in 2000 with the ambitious target of eradicating fuel poverty, provides means-tested grants of up to £3,500 (or £6,000 where oil central heating is an option) to provide poorly insulated homes with insulating energy-efficiency solutions. Over two million homes have received grants for improvements between 2000 and 2009.[9] However, the scheme was recently cutback by the Coalition government, with a smaller fund of £210 million available for the period 2011 to 2013 and stricter eligibility criteria.[10]

Warm Front has been quite successful in combating fuel poverty. According to DECC and Eaga, the lead contractors, it achieved average savings of £650 for every household insulated in 2009/10.[11] Return on the investment was not immediate but substantial, in that £852 million of expenditure from 2005 to 2008 yielded savings for the same period of over £240 million.[12] The contractors also found a high level of customer satisfaction, 86 per cent, with the work done. Expressions of dissatisfaction generally concerned customers asked to contribute some of the costs themselves or arrange their own boxing in of pipes or repairs to plasterwork.

Currently, Warm Front overlaps to some extent with other policies, undermining its overall effectiveness. The Carbon Emission Reduction Target has particular goals to improve energy efficiency for vulnerable groups and it has led to energy companies contributing funds to the Warm Front scheme to discharge the obligation for them. The Community Energy Saving Programme (CESP), a wide-ranging collection of community-based schemes, includes a scheme called Warm Zones with similar objectives. A quick examination of this description from DECC's 2009 CESP consultation document suggests that insulation courtesy of Warm Front could achieve just as much by itself:

> Warm Zones is an area-based programme that aims to address fuel poverty. It integrates funds and grants from a wide range of sources, co-ordinating the delivery of benefits that include: energy efficiency, carbon savings, fuel poverty reductions, income maximisation, health improvements, fire and home security, employment skills and training.

> Warm Zones works right across the housing sector, aiming to deliver 'something for everybody'. This means accessing the hard-to-reach and vulnerable households at greatest risk from the adverse impacts of fuel poverty.[13]

It would be more straightforward to leave all domestic insulation and energy-efficiency schemes to an enlarged Warm Front programme.

While eradicating fuel poverty is a worthy objective, it makes even greater sense, when prices are being raised artificially to fund research for a lower-cost future. Of the three causes of fuel poverty, namely energy-inefficiency, high fuel prices and low incomes, Warm Front needs to be able to address every case affected by the first. As with any subsidy, a small amount of equalisation will happen when the scheme is in operation: fuel prices will rise a fraction as the market takes into account people's increased ability to pay. This is something for the regulator to regulate. As for the third factor, in the vast majority of cases, when incomes are so low that even heating an energy-efficient home requires more than a tenth, other schemes are needed besides. For a household that poor, it is likely that fuel poverty will not be the only form of poverty to which they are exposed.

To ensure that all energy-inefficient homes are eligible requires a new look at means-testing. Means-testing often discourages people who are put off by intrusive and time consuming form-filling from applying and also rules out many people who could make substantial savings. The National Audit Office said of the means-testing:

> Analysis of the English House Condition Survey 2006 indicates that 57 per cent of vulnerable households in fuel poverty do not claim the relevant benefits to qualify for the Scheme. Yet nearly 75 per cent of households who would qualify were not necessarily in fuel poverty.[14]

It is preferable to include some people who could fund improvements themselves instead of excluding people without resources. To this end, means testing should only apply for the early period of the scheme or for the first six months of each year. After that any remaining funds should be open for anyone to apply who could meet a surveyor's criteria for worthwhile improvements. The worst potential abuse would be that a wealthy citizen acquired a large draughty property and had it insulated at the scheme's expense. Placing an upper limit on pay-outs for individual grants could prevent this. More importantly, every household in the most vulnerable category is rendered eligible and no-one has to admit to poverty in order to join the scheme. Even grants to wealthy people have the benefit of improvements to the housing stock.

Taking 2009 levels of fuel poverty and adjusting them to allow for immediate increases in price for a £30 carbon tax, a crude estimate is that £7.5 billion of Warm Front expenditure would eradicate it. Some will persist, due to absolute poverty or inevitable mis-targeting, but that order of expenditure has the potential to transform it into a rare form of poverty, at least until further rises in fossil fuel price, independent of climate change regulation, raise the price again. No apology is necessary for increasing the demand for British-made insulation panels, reducing emissions and improving the security of energy-supply.

The carbon tax at £30 would, if paid in full, yield £7 billion annually from installations registered under the EU ETS scheme and a further £2 billion from domestic gas users. Domestic electricity users will pay the tax from the start, wherever it is passed through by the generators, as expected in most cases. A range of choices is possible about the timing of grants for research and to Warm Front, and when to introduce the tax. An attractive option is to delay the tax on domestic gas for three years, giving time for people to take up grants for improved insulation. This would free up the revenues from the carbon tax in subsequent years to be spent on subsidising research and development into low-cost, low-carbon technologies.

Investment spend

The proposed carbon tax will cause an immediate increase in fuel bills that then remains stable, subject only to the fluctuations in basic fuel prices. Admittedly, those fluctuations could be substantial but they are no worse than the market causes already. Because those increases will have an adverse effect on people deemed fuel poor, it is proposed that a large part of the early tax revenues are spend on alleviating fuel poverty.

Calculating the scale of investment needed relies on several assumptions. Recent Warm Front results suggest that annual savings amount to around 40 per cent of the investment.[15] For simplicity, it may be assumed that all households may be improved in this manner, although this is an over-simplification, since there are some cases where the relevant technologies are not applicable. Likewise, it is optimistic to expect all investment to reduce bills to the cut-off point for fuel poverty and no further. It is worth remembering that any improvement in energy efficiency brings immediate benefits to household disposable incomes, reductions in emissions and improvements in energy security.

Since fuel poverty is a theoretical concept, it is possible to know exactly the level of notional expenditure at which an average household enters fuel poverty, so the investment to reduce fuel bills to that level may be calculated. As the tax increases prices, so some extra investment is necessary for further reductions for households already in fuel poverty, an estimate for people only rendered fuel-poor by the tax-related increase.

The end result is quite a loose estimate, but it shows the scale of what can be achieved. If the carbon tax brings in £9 billion in its first year, £5 billion would remove existing poverty, under the generous assumptions described. To remove further effects of the tax, £0.5 billion would, assumptions permitting, prevent any additional households falling into the category. £2 billion more would bring the existing fuel poor to an improved efficiency standard that would keep their bills below the threshold even with the tax. Discounting continued commitments, this would leave £0.8 billion for LCLC investment in the first year of the tax. The basic point of these estimates of scale is that a single year's intensive investment can relegate fuel poverty from a headline to a side-issue. Thereafter, the tax may be spent partly on existing obligations under schemes to subsidize renewable, but mostly on research to bring down the price of low-carbon energy. However, the spending allocation could be balanced out according to political need, and it would perhaps be better to spend equally on LCLC research and fuel poverty, to improve both over a steady time period without encouraging the waste of resources.

These calculations are based on DECC 2009 fuel poverty statistics, taking the saving required to reduce fuel expenditure to ten per cent of income.[16] The effect of the carbon tax on bills assumes typical emissions for each pound spent at current prices. Some of these assumptions may be over-optimistic. Yet there are enough stories of energy-inefficiency to suggest that large savings are possible within households.

However as the incentive effects of a carbon tax take hold and companies, households and industries seek less carbon-intensive energy sources, the revenues from the carbon tax will begin to diminish. This should not mean that the rate of the tax should increase to compensate as this would cause a loss in investor confidence and undermine one of the main advantages of a tax over a trading scheme, that of certainty. Indeed, the huge injection of funds into LCLC research and the inception of the tax would mean that this declining income would not necessarily be a problem.

Tariffs and committed costs

There are some costs to be considered if all existing environmental charges were dropped in favour of the flat carbon tax. For example, while it might be reasonable to scrap the feed-in tariffs, it would be less so if the government stopped paying out the revenue it promised to those who had set up renewable power generation under the assumption that they would continue to be paid for years to come. The burden of these committed costs is best dealt with by paying out from the revenues of the carbon tax, so our calculations generating the true amount to be spent on LCLC R&D and fuel poverty must take these into account.

Renewables obligation

The Renewables Obligation (RO) is designed to incentivise the generation of electricity from eligible renewable sources and it obliges large electricity suppliers to source a percentage of their electricity from these, with a fine if they do not. Put simply, for every megawatt hour (MWh) generated through renewables, a certificate is given to the supplier, known as a Renewables Obligation Certificate (ROC). If the supplier does not have enough ROCs to cover the required percentage of output, they can pay a 'buy-out' price. This currently stands at £38.69 per MWh.[17] If they invest to avoid the fines, they will gain only the freedom from the fines for the remaining duration of the scheme. A carbon tax on all carbon-based electricity production is effectively a 100 per cent renewables obligation, although nuclear energy would escape the tax too. There is no need to recompense holders of ROCs for the abandonment of the scheme unless the difference in price is marked, and possibly not even then.

£37 per MWh equates at current averages for electricity generation to approximately £70 per tonne of saved CO_2. If the carbon tax is set at £30, there is a slight case for recompensing generators for the difference of £40 per MWh. As mentioned above, if generators fail to submit their quota of ROCs, then they pay fines, and the revenue from this is distributed to ROC holders. Since these payments are conditional on other generators' failure to comply, there is never any guarantee than they will happen, and therefore there is no need to provide compensation for ending this aspect of the scheme. This may be subject to legal challenge.

In John Constable's *The Green Mirage*, it is assumed that the annual unit price for a ROC is £50.[18] That allows also for the payments from non-compliers to compliers, which this report hopes to avoid by ending the scheme quickly. Based on ROC returns for 21 million MWh in 2009/10, compensation at the rate of £40 per tonne or £22 per MWh would cost nearly £500 million, which would have to be found from the carbon tax proceeds before they could be more usefully deployed. It is important to note that the longer the scheme runs without reform, the higher these potential compensation sums will become. It is therefore cheaper to move to a carbon tax now, rather than wait in the hope for the EU ETS rectifies itself by 2016-18 and have to switch later.

Feed-In Tariffs

Feed-In Tariffs for microgenerators of renewable energy attract much higher subsidies than anywhere else in the power sector. Presently, only 108MW of capacity have been registered to take advantage of Feed-In Tariffs, which is tiny by comparison with the schemes under the Renewables Obligation.[19] However, a tariff at the higher, most attractive end of those available in 2011 is 43.3p per kWh, representing a huge £433 per MWh, or £730 per tonne of saved fossil-fuel-generated CO_2.[20]

Microgenerators are in many cases private citizens, who have taken investment decisions, often for environmental as well as financial reasons. Denying them the subsidies would risk discrediting any government scheme to promote environmental responsibility. The subsidies already committed to therefore need to continue to be paid.

Assuming similar rates of utilisation to the RO scheme, the registered Feed-In Tariff installations could produce around 300 GWh annually. This is a small part of overall national capacity but still commands an annual subsidy in the region of £130 million. This is not enough by itself really to lessen the funds available through the carbon tax and many installations receive lower rates of subsidy anyway, so this figure is probably at the upper end of reasonable estimates. Nevertheless, subsidy above £1 million per MW of capacity is an extremely expensive way of saving emissions. Unless the scheme is reformed or ended quickly, there is potential to incur a billion pounds of annual liability in return for under a GW of generating capacity.

As an aside, a grievance with the Feed-in Tariff regime is that it pays for quantities generated, even when they are immediately consumed by the generator and do not supply the National Grid. With rates that high, it is possible to imagine abuses whereby sites could turn on lights over solar panels and receive back more for the generated solar power than they paid for the electricity. A happier alternative is to pay the subsidy only for energy exported to the National Grid, thereby available to benefit wider society. A unit saved is as valuable in reducing emissions as a unit generated renewably.

Finally, on the subject of domestic tariffs, it is an anachronism for energy-users to be charged less for their usage above a threshold than for the basic usage below it. It is a vestige of the 'standing charges' that applied to being connected at all. In circumstances of scarce resources, prices need to allow for a measure of cheap, basic usage, covering unavoidable fuel use for civilized living conditions. The higher price belongs above the threshold, giving households greater benefit from any savings they can engineer. When the same households are also generating their own power, each saved unit can earn the saving at the higher unit rate.

APPENDIX 3

An Installation and its Allocations – Understanding the ETS in terms of One Company's Behaviour

The installations in this illustration are hypothetical and do not represent any particular site. It is intended to show the practical effects of its various measures. The size chosen, emitting 30,000 tonnes of CO_2 per annum, represents an emitter around the median but a lot smaller than the largest emitters in the scheme. The same principles apply to installations of any size.

Phase I

Before the scheme began, the company had replaced a coke-fired furnace with a gas one at the same time as reducing capacity by a quarter. On the basis of previous usage, it was allocated credits worth 50,000 tCO_2e per annum. It therefore enjoyed credits of 20,000 tCO_2e per annum, received free but surplus to what it needed for compliance. Phase I saw no further change in emissions.

Because of the general over-supply, attempt to sell spare units did not raise an attractive price. The company sold 15,000 tCO_2e profitably in each of the first two years of Phase I, 2005 and 2006, but when selling in 2007, after the price had fallen to near zero, the company gave up trading after getting very little return for the first 10,000 tCO_2e. After three years, it was not able to carry forward any surplus to Phase II and the remaining 20,000 tCO_2e not needed to cover emissions were lost.

Phase II

In Phase II, certain extra categories of emission were included in the allocations. Because our example company produced some gypsum in the course of its business, a second installation was brought into the scheme and its annual emissions within the scheme rose by 3,000 tCO_2e to 33,000 tCO_2e. After some renegotiation of the UK's National Allocation Plan, the company's annual allowance in Phase II was cut to 45,000 tCO_2e. The annual windfall became 12,000 credits per annum.

At this stage in the scheme, some installations had emerged whose free allocations did not meet their emissions requirements. Our example company managed to sell their annual surplus for three years, making a windfall profit for themselves. By the close of 2010 they had cashed 36,000 credits and retained an annual expected surplus of 12,000.

They also began to enjoy a competitive edge over less fortunate competitors who, required to pay for some of their emissions permits, were obliged to pass costs they could not absorb to their customers. Our example company was able to take advantage by a mixture of slightly lower prices and greater profit margins.

At this point, Phase III was two years distant, with a prospect of tighter controls and higher prices. Not being in the power sector, the example company escaped from any need to buy their allowances at auction. Permitted to meet 20 per cent of their emissions with CERs, the company bought 20 per cent of its 33,000 required credits each year, retaining more of their home allocation. For 2011 and 2012, they had 45,000 plus 6,600, making 51,600. In order to fund the purchase of CERs, they sold a lesser number of European credits to other installations that were unable to meet their deficits by CERs alone. Selling 5,600 each year left them with 46,000, a surplus of 13,000 over what they needed for their own emissions.

With some of the windfall money, they expanded the plant in 2011, re-establishing some of the market share lost just before the scheme began. Annual emissions rose from 33,000 to 40,000 tCO2e. This reduced their annual surplus to 6,000 tCO_2e. At the close of Phase II in December 2012, they would have 12,000 tCO_2e in reserve and in net terms have paid nothing for their allowances.

Phase III

In Phase III, changes may begin to bite. The allowance reduces by 1.74 per cent each year. It would take 12 years before the allowance would reduce as far as the installations' actual emissions. Some sectors will still receive free allowances in Phase III. As an energy-intensive company, our example is saved from auctioning and continues to be awarded its allowances free of charge.

Even though, by extremely fortunate timing, it had enjoyed a cushion from any need to make reductions through Phases I and II, Phase III will be tougher. Benchmarking may reduce the allowance from 45,000 to 28,000 tCO_2e, which may be more typical of companies that have not had the benefit of windfalls. The shortfall in the allowance will be 5,000 in the first year, but with a reserve from Phase II of 12,000. For the first time, the example installation has to choose between making actual reductions in emissions and using up its surplus. After that, a free allowance of 28,000 tCO_2e becomes 27,500, after a reduction of 1.74 per cent. Needing to find so many credits each year, the accumulated surplus of 12,000 will be spent before three years are up.

As an alternative, they may buy CERs up to half the required reduction below 2005 levels.[21] If that means 2005 allocations, then the required reduction is the difference between 50,000 and 28,000, and half of that is 11,000 tCO_2e. That is considerably more than the company's annual shortfall. If permitted, that would allow them to carry on buying CERs cheaply and selling ETS credits at a higher price. The precise rules are very hard to interpret, so it is unlikely that any installation would be permitted to use quite such a substantial quantity of credits from outside the EU.[22] Buying even 6,000 CERs over two years would extend by a year the time taken to use up the company's accumulated surplus. After that, the example company will have to pay for the privilege of emitting greenhouse gases, unless some aspect of the scheme changes before then. By that time, ETS will have had ten years of continuous operation.

It should be stressed that this is an imaginary but not impossible example. Many companies will have had more difficult experiences. Even with a less generous initial allocation, a company may bank cheaper, non-European credits throughout Phase II and only contemplate serious emissions reductions when required to pay at auction for half the allowance from Phase III. The fact that it is possible for a scheme to run for ten years, while some participants take no steps towards its stated aims and have the chance to make money at customers' and competitors' expense, suggests that it is far from an ideal policy.

APPENDIX 4

The Glut Graph

The data for this graph are drawn from many sources and every effort has been made to retain compatibility. For each phase, the scheme total is divided between national allocation plans. For Phases I and II, countries divide their totals between installations. Emissions are expected to remain within allocation. They may be carried over during Phases I and II and the early part of III. Estimates of allocations to installations for 2011 are by extrapolation, continuing the linear trend from 2008 to 2010. Estimates of allocations to installations for 2012 are on the basis of spending the remaining cap from national Phase II allocations. The figure given is from January 2011. The Phase III cap includes the non-EU members, particularly Norway. The graph omits allowances for Romania and Bulgaria in order to preserve continuity.

Estimates for Norway's Cap (Liechtenstein's too) are derived from Defra tables. These give the 2009 allocation installations and the headroom ('length') compared to both the cap and the allocation to installations. The cap is the allocation to installations, plus the difference between these two measures of length. Some reports treat auctioned allowances differently from free allocations.

References:

Phase I and II caps:
http://europa.eu/rapid/pressReleasesAction.do?reference=IP/07/1614

Phase II cap (January 2011): http://ec.europa.eu/clima/policies/ets/cap_en.htm

Emissions and Allocations Installations Data:
http://dataservice.eea.europa.eu/PivotApp/pivot.aspx?pivotid=473

Norway Cap calculations:
http://publications.environment-agency.gov.uk/PDF/GEHO1010BTDK-E-E.pdf

Country conventions:
http://www.eea.europa.eu/data-and-maps/data/european-union-emissions-trading-scheme-eu-ets-data-from-citl-3/eu-ets-data-viewer-manual/eu-ets-data-viewer-manual/at_download/file

APPENDIX 5

No Need to Act on Emissions until 2016-18

Headroom, the difference between allocations and emissions, is critical to the functioning of the market. If the market is to work to reduce emissions, there should not be any spare at all. Because there has been surplus through much of Phase II, it is expected that a substantial volume of credits will be carried over to Phase III. Estimates of how long the surplus will last depend on projections of future emissions. A small change in projected emissions can lead to a large change in the time taken to use up the surplus. The change constitutes a larger proportion of the surplus than of the emissions. What follows builds estimates upon estimates but describes a future situation wholly within the parameters of the ETS as presently constituted.

Figure 1:1 has assumed that all national allocations are passed to installations at the end of Phase II, and that the headroom at the start of Phase III is the difference between those allocations and the emissions that year. As the cap reduces in Phase III, that headroom will be gradually used up, or so one might hope. Not until 2016 does the cap fall to the level of 2010 emissions. If emissions remain at that level, determined in part by an incomplete recovery from recession, the carryover from Phase II can remain in the bank until 2016, and will not all be used up by 2020. If other things remain equal, which they seldom do, the falling cap of Phase III applies no pressure to fall below 2010 levels of emissions until after 2020.

It is worth looking at which things may not remain equal. It is possible that emissions will rise during that period. In that case, the pressure to reduce emissions may come earlier in the decade but it will come from rising emissions. The incentive to reduce greenhouse gases derives from greater release of greenhouse gases. That is not an incentive worth encouraging.

Other differences include the potential to include more categories of emissions within the cap. Aviation is the most obvious. If airlines are competing for the same permits as everybody else, the time when the spare from Phase II is used up will arrive several years earlier. Airlines are in fact granted a separate, additional allocation in Phase III, but other increases in scope amounting to around 108 million credits annually will accelerate using up the spare credits. Even with the increase in scope, the Phase III caps and the Phase II carryover are enough to maintain emissions at 2010 levels until 2017.

Then there is the gradual move to auctioning permits, rather than giving them away. This applies little extra pressure to reduce emissions. The windfall element is shifted from the installations to the national governments. Installations will not be able to acquire cheap spare permits unless there is nobody else wishing to buy them. Even so, unless there is an excess of demand over supply, prices bid at auction need not be high. If there is spare capacity, a single bid at any price will be enough to secure permits for an installation. Auctioning still depends on having no remaining headroom within the system.

There may be an extent to which installations should be rewarded for reducing their emissions early. If 2010 levels represent an impressive achievement, there may be a case for rewarding installations for bringing down their emissions so quickly. The difficulty is that the cap remains high, and has been rendered higher by the addition of credits from outside Europe, many of which have not been environmentally beneficial. The reward for the early reducer depends on someone later on wishing to pay for the right to the saved emissions. As long as enough permits exist to go round, the price of the reward remains low. Every further saving in emissions puts off the time when the installation making the economy can profit *via* the trading scheme. Efficiency savings will profit the company that finds them straight away, but the bonus from the ETS will not arrive until the receding time when demand for permits exceeds supply. Futures markets do allow a means whereby a permit can have some value today although its intrinsic value relates to a future time. On current projections, that future time is not until 2017.

APPENDIX 6

How to Account for Nuclear Capital Costs

The most complete source of recent comparative cost data is the Mott MacDonald study for DECC, which uses a variety of scenarios and assumptions.[23] In most cases, the cheapest low-carbon option is onshore wind, but the important comparison is between nuclear and offshore wind, since those already have the potential for large-scale low-carbon generation. Running costs are generally lower for nuclear, so it depends on the capital situation. For nuclear stations, the capital costs depend on front-loaded construction costs and end-loaded decommissioning costs. The accounting convention of discounted cash flows means that certain assumptions can have a surprising effect.

Mott MacDonald plans a 25 year life for a wind farm but 60 years for a pressurized water reactor. They have a base-case discount rate of 10 per cent, meaning that a cost may be that much greater if it is a year longer before it has to be spent. For a cost of £1 in the first year or a cost of £1.10 in the second year are equivalent.

The Mott MacDonald report lists costs after applying discounting so it requires some arithmetic to see what figures would look like under different assumptions:

First, life of the plant. Repaying capital costs at the rate of £49.20/MWh over 60 years could be achieved in 25 years at the rate of £54.02. This would reduce the nuclear advantage over one version of offshore wind from £44 to £39 per MWh but the advantage would still be in that direction.

Second, discount rate. A lower discount rate, as shown in Mott MacDonald's Case 8 relative to their Case 6, would further reduce the gap, as offshore wind has greater front-loaded capital costs than even nuclear. Removing the discount rate altogether brings them closest over a 25-year lifetime. That is not usual accounting or investment practice but it can be helpful for a scheme to be economically advantageous in cash terms, regardless of when the money is paid or received. The gap in favour of nuclear generation remains £12 per MWh over a 25 year life for capital repayment and a further £5 in lower running costs.

Decommissioning is more complicated. It is incorporated in the projections as a provision of £2.10 per MWh over a 60 year life. Most of those costs will occur at the end of that lifetime. If those costs use the same discount rate, their combined discounted value is several times that of the initial capital required to build the power station. It is mildly noteworthy that costs are stated as the same whether the discount rate is 10 per cent or 7.5, although there appears to a substantial gap between the time the provision is made and the time the money is spent. Plans published for specific sites by the Nuclear Decommissioning Authority use a much more cautious discount rate of 2.2 per cent, which prevent large costs from looking lower in the far distance.[24] As estimates based on these plans conclude that the decommissioning process, though expensive, costs less than building the station, there are grounds for believing they have not been under-estimated, at least no more so than the costs of building any other form of power-generation.[25]

Overall it is the capital costs, whether related to construction prior to generation or decommissioning after it, where the savings must be found to make nuclear power the low-carbon alternative, preferable to gas and coal economically as well as environmentally. Using research to find less costly processes is preferable to using favourable accounting conventions.

APPENDIX 7

The Emissions Saved from Switching from Coal to Gas

This is a look at the suggestion that all coal-fired power-generation in the UK could be replaced by gas-fired generation. The annual saving in CO_2 emissions, based on 2010 statistics, would be around 34 $MtCO_2e$.

Tables in the DECC DUKES report (the Digest of United Kingdom Energy Statistics) reveal that in 2010 UK power generation ran to 370,000 GWh of gas burnt to make electricity[26] and 41.5 million tonnes of coal, equivalent to around 340,000 GWh of energy.[27] Replacing the coal fuel with gas, but keeping the same levels of efficiency in the rest of the system, there is a potential saving because gas produces less carbon dioxide for the same energy. It is possible that newer turbines could increase the efficiency further but that has not been included here, as any technology could benefit as much from assumed increases in efficiency. The difference in emissions per GWh is approximately 100 tCO_2e. Applying that to the entirety of coal-fired generation yields annual savings of 34 $MtCO_2e$. UK CO_2 emissions in 1990 were 585 $MtCO_2e$, so this represents a potential further 5.8 per cent saving on what has already been achieved.[28]

According to the same DUKES report, 340,000 GWh represents 31 per cent of current gas demand. Replacing coal-fired generation with gas would add approximately a third to UK demand, with consequences for reserves, availability and prices.

Select Bibliography

Constable, J., *The Green Mirage: why a low-carbon economy may be further off than we think*, London: Civitas, July 2011

Ellerman, A., Convery, F., & de Perthuis, C., *Pricing Carbon,* Cambridge: Cambridge University Press, 2010

Environmental Audit Committee, *The role of carbon markets in preventing dangerous climate change,* February 2010

Environmental Audit Committee, *Carbon Budgets, Third report of the session 2009-10,* Volume 1, January 2010

Helm, D., 'Green growth: opportunities, challenges and costs' in Tsoukalis, L., & Emmanouilidis, J., eds. *The Delphic Oracle on Europe: Is there a Future for the European Union?*, Oxford: Oxford University Press, May 2011

Helm, D. & Hepburn, C., (eds), *The Economics and Politics of Climate Change*, Oxford: Oxford University Press, 2009

House of Common Library, *EU ETS and Aviation*, SN/SC/5533, April 2011

MacKay, D., *Sustainable Energy – Without the Hot Air*, Cambridge: UIT, September 2010

Merlin-Jones, D., *Chain Reactions: How the Chemical Industry Can Shrink Our Carbon Footprint*, London: Civitas, June 2011

Mott MacDonald, *UK Electricity Generation Costs Update,* June 2010

Prins, G., *et al, The Hartwell Paper: A new direction for climate policy after the crash of 2009*, May 2010

Sandbag, *Carbon Fat Cats 2011*, June 2011

Sandbag, *Emissions Trading in the EU: Why it is still not working and some ideas for what can be done about it,* July 2008

Sandbag, *Rescuing the EU ETS from redundancy*, April 2010

World Bank, *State and Trends of the Carbon Market 2010*, May 2010 & *2011*, June 2011

Europa website (www.europa.eu)

Department of Energy and Climate Change website (www.decc.gov.uk)

Notes

Prelims

1 The Royal Society, *Economic instruments for the reduction of carbon dioxide emissions*, Policy Document 26/02, November 2002,

2 Sowell, T., *The Vision of the Anointed – Self Congratulation as a Basis for Social Policy*, Basic Books, 1995

3 The 'S' in EU ETS is sometimes interpreted as 'Scheme', a term used in EU documents, although it is most commonly described as a 'System' by DECC and the European Commission. Given that there is no definitive right term, this report refers to it primarily as the Emissions Trading System.

4 http://ec.europa.eu/clima/faq/ets/index_en.htm, Paragraph 1

5 http://dataservice.eea.europa.eu/PivotApp/pivot.aspx?pivotid=475

6 http://europa.eu/rapid/pressReleasesAction.do?reference=IP/08/787&format=HTML&aged=0&language=EN&guiLanguage=en

7 http://europa.eu/rapid/pressReleasesAction.do?reference=IP/07/1614&format=HTML&language=EN#fn12

8 http://www.decc.gov.uk/en/content/cms/what_we_do/change_energy/tackling_clima/emissions/eu_ets/euets_phase_ii/auctioning/auctioning.aspx

9 http://www.decc.gov.uk/en/content/cms/what_we_do/change_energy/tackling_clima/emissions/eu_ets/phase_iii/phase_iii.aspx

10 http://europa.eu/rapid/pressReleasesAction.do?reference=IP/03/931&format=HTML&aged=0&language=EN&guiLanguage=en

11 Environmental Audit Committee, *The role of carbon markets in preventing dangerous climate change,* February 2010, 5.94

Chapter 1

1 *The role of carbon markets*, 3.17-20

2 Ellerman, A., Convery, F., & de Perthuis, C., *Pricing Carbon,* Cambridge University Press, 2010, p. 181

3 http://uk.reuters.com/article/2011/02/02/greenbiz-us-carbon-investment-idUKTRE7115K520110202

4 Sandbag, *Emissions Trading in the EU: Why it is still not working and some ideas for what can be done about it*, July 2008, pp. 8-9

5 *Pricing Carbon*, p. 56

6 http://europa.eu/rapid/pressReleasesAction.do?reference=IP/07/1614

7 http://europa.eu/rapid/pressReleasesAction.do?reference=IP/07/1614

8 http://www.environmental-finance.com/news/view/334

9 From EU Environmental Agency data. German total CER credits of 83,095,436 and ERU credits of 4,865,496 with a total cap of 453.1 $MtCO_2e$ according to http://europa.eu/rapid/pressReleasesAction.do?reference=IP/07/1614&format=HTML&language=EN#fn5

10 World Bank, *State and Trends of the Carbon Market 2010*, May 2010, p. 57

11 http://www.votewatch.eu/cx_vote_details.php?id_act=2043&lang=en

12 Committee on the Environment, Public Health and Food Safety, *On the analysis of options to move beyond 20% greenhouse gas emission reductions and assessing the risk of carbon leakage*, (2011/2012(INI)), June 2011, paragraph 19

13 http://www.reuters.com/article/2011/07/01/us-barcap-carbon-idUSTRE76015T20110701

14 Environmental Audit Committee, *Carbon Budgets, Third report of the session 2009-10*, Volume 1, January 2010, p. 4

15 Merlin-Jones, D., *Chain Reactions: How the Chemical Industry Can Shrink Our Carbon Footprint*, London: Civitas, June 2011, p. 28

16 Carbon Trust, *Raising the Bar*, July 2011, p. 1

17 National Audit Office, *European Union Emissions Trading Scheme*, p.8

18 *Raising the Bar*, p. 3

19 Point Carbon, *Carbon 2010: Return of the Sovereign*, 2010, p. 19

20 Sandbag, *Rescuing the EU ETS from redundancy*, April 2010, pp. 1-2

21 *Rescuing the EU ETS*, p. 2

22 *Rescuing the EU ETS*, p. 4

23 Sandbag, *Carbon Fat Cats 2011*, June 2011, pp. 6-14

24 CE Delft, *Will energy-intensive companies profit from EU ETS under Phase 3?*, Delft, December 2010, pp. 40-42

25 Sandbag, *Buckle Up! Tighten the cap and avoid the carbon crash*, July 2011, p. 7

26 Estimates based on a figure from Point Carbon, *EU ETS Phase II – the potential and scale of windfall profits in the power sector*, March 2008, p. 2 The original figures of €23 to €71 billion were based on an out-dated EUA price estimate of €21 to €32 during Phase II. This report uses an updated average figure of €13.6 per EUA to reflect the fall in their price.

27 *The role of carbon markets*, 4.15

28 *The role of carbon markets*, 3.21

Chapter 2

1 http://ec.europa.eu/clima/policies/ets/linking_en.htm

2 Centreforum, *Improving the EU emissions trading scheme,* August 2008, p. 7

3 http://www.businessgreen.com/bg/news/2047135/businesses-quickly-discard-banned-carbon-credits

4 European Environment Agency, CITL database

5 http://www.reuters.com/article/2011/06/16/us-offsets-point-carbon-idUSTRE75F3A720110616

6 http://www.climate-connect.co.uk/Home/?q=node/742

7 Öko-Institut, *2010 Rating of Designated Operational Entities (DOEs) accredited under the Clean Development Mechanism (CDM)*, May 2010, p. 4

8 Schapiro, M., 'Conning the Climate: Inside the Carbon-Trading Shell Game', *Harper's*, February 2010, p. 36

9 http://www.co2offsetresearch.org/PDF/AdditionalityLackCDM.pdf

10 http://www.guardian.co.uk/environment/2008/may/21/environment.carbontrading

11 IPCC & TEAP, *Special Report on Safeguarding the Ozone Layer and the Global Climate System: Issues Related to Hydrofluorocarbons And Perfluorocarbons*, 2005

12 http://www.eia-international.org/cgi/news/news.cgi?t=template&a=626&source=

13 Sandbag, *Industrial Gas Big Spenders: HFC and N20 adipic credit usage in 2010*, May 2011

14 CDM Watch & EIA, *HFC-23 Offsets in the Context of the EU Emissions Trading Scheme*, July 2010

15 http://www.guardian.co.uk/environment/2010/oct/26/eu-ban-carbon-permits

16 Carbon Trade Watch, *EU Emissions Trading System: failing at the third attempt*, April 2011, p. 3

17 *Industrial Gas Big Spenders*

18 Point Carbon from http://www.rumoursandfacts.com/2011/09/21/china-clears-record-batch-of-cdm-projects/

19 http://www.businessgreen.com/bg/news/2031480/banned-industrial-gas-credits-sneak-eu-carbon-market & *HFC-23 Offsets in the Context of the EU Emissions Trading Scheme*

20 http://cdmpipeline.org/regions_7.htm

21 Directive 2009/29/EC of the European Parliament and of the Council, April 2009, paragraph 30

22 *Emissions Trading in the EU*, p. 13

23 *The role of carbon markets*, 3.38

Chapter 3

1 http://www.bbc.co.uk/news/business-14151032

2 http://blog.decc.gov.uk/2011/07/28/the-true-cost-of-energy-and-climate-change-policies-on-bills/

3 Ofgem, *Updated Household Energy Bills explained*, August 2009

4 Ellerman, A .& Loskow, P., *The European Union's Emissions Trading Scheme in perspective*, May 2008, p. 27

5 http://www.utilityweek.co.uk/news/news_story.asp?id=195616&title=Next+phase+of+carbon+trading+will+hike+Europe%27s+power+prices+by+up+to+20+per+cent

6 European Commission, *Europeans' attitudes towards climate change*, September 2008, pp. 6 & 68

7 http://www.epcengineer.com/news/post/5396/centrica-worried-about-support-for-uk-energy-policy-goals

8 http://www.decc.gov.uk/en/content/cms/what_we_do/change_energy/tackling_clima/emissions/eu_ets/euets_phase_ii/auctioning/auctioning.aspx

9 DECC, *Guidance for indirect bidders*, October 2008, updated June 2009

10 DECC, *FAQs on competitive auctioning, updated May 2010*, May 2010, p. 3

11 DECC, *Summary of feedback on first UK auction in Phase II of the EU Emissions Trading System*, January 2009, pp. 2-3

12 Friends of the Earth, *Ten Ways to Game the Carbon Market*, May 2010, p. 7

13 Bradsher, K., 'Outsize profits, and questions, in effort to cut warming gases', *New York Times*, 21 December 2006

14 http://wbcarbonfinance.org/Router.cfm?Page=UCF

15 http://www.reuters.com/article/2010/08/23/us-hfc-bank-hfc-idUSTRE67M2CK20100823

16 http://www.europol.europa.eu/content/press/further-investigations-vat-fraud-linked-carbon-emissions-trading-system-641

17 Reproduced in full from World Bank, *State and Trends of the Carbon Market 2011*, June 2011, p. 40

18 *State and Trends of the Carbon Market 2011*, June 2011, p. 44

19 http://ec.europa.eu/clima/news/articles/news_2011032201_en.htm

20 http://climateandcapitalism.com/?p=4363

21 http://www.europol.europa.eu/content/press/further-investigations-vat-fraud-linked-carbon-emissions-trading-system-641

22 http://www.europol.europa.eu/index.asp?page=news&news=pr091209.htm

23 http://www.telegraph.co.uk/finance/newsbysector/energy/8290533/The-great-carbon-trading-scandal.html

24 Carbon Trust, *Cutting Carbon in Europe: The 2020 plans and the future of the EU ETS*, June 2008, p 23

25 ONS, HM Revenue and Customs receipts, Total HMRC, May 2011

26 EU Official Journal, *Directive 2008/101/EC of the European Parliament and of the Council, amending Directive 2003/87/EC so as to include aviation activities in the scheme for greenhouse gas emission allowance trading within the Community*, January 2009, Article 3d.4

27 DECC, Consultation on the EU Commission's proposals to amend the EU ETS from 2013, May 2008

28 DECC, *FAQs on competitive auctioning*, p. 1

Chapter 4

1 *Chain Reactions*, p. 56

2 Aldersgate Group, *Trading for Growth: The role of the EU ETS in cutting emissions and stimulating wealth creation*, November 2007, p. 5

3 UK Trade & Investment, *Chemicals, the UK Advantage*, April 2009, p. 3

4 *Chain Reactions*, pp. 5-6

5 Waters Wye Associates, *The Cumulative Impact of Climate Change Policies on UK Energy Intensive Industries – Update Against New Government Policy*, March 2011, p. 3

6 *The Cumulative Impact of Climate Change Policies on UK Energy Intensive Industries*, p. 3

7 *The role of carbon markets*, 3.48

8 http://www.steelguru.com/sfTCPDF/getPDF/MTM0NjU0/Downsizing_deals_-_Corus_to_receive_GBP_250_million_in_carbon_credits.html

9 http://www.guardian.co.uk/business/2011/feb/24/corus-ssi-sale-teesside-jobs

10 Helm, D., 'Green growth: opportunities, challenges and costs' in Tsoukalis, L., & Emmanouilidis, J., eds. *The Delphic Oracle on Europe: Is there a Future for the European Union?*, Oxford University Press, May 2011, p. 17

11 Environmental Audit Committee, *Memorandum submitted by David Newbery, Research Director, Electric Policy Research Group, University of Cambridge*, February 2009, Paragraph 3

12 House of Common Library, *EU ETS and Aviation*, SN/SC/5533, April 2011, p. 7

13 Only light aircraft are excluded from the EU ETS, for the sake of regulatory simplicity.

14 http://www.oilandgasobserver.com/commentary/aviation-discord-persists-ahead-of-implementation-of-the-eu-emissions-trading-scheme/003977

15 Standard & Poor's, *Airline Carbon Costs Take Off as EU Emissions Regulation Reach For The Skies*, February 2011, p. 2

16 Ernst & Young and York Aviation, *Analysis of the EC proposal to include aviation activities in the Emissions Trading Scheme*, June 2007, p. 33

17 Europa, Questions & Answers on the benchmark for free allocation to airlines and on the inclusion of aviation in the EU ETS, 26/09/2011

18 http://ictsd.org/i/news/biores/109559/

19 http://euobserver.com/884/32316

20 http://www.prnewswire.com/news-releases/air-transport-association-calls-eu-ets-application-to-us-airlines-illegal-125018899.html

21 http://www.ft.com/cms/s/0/ef1e95c0-e097-11e0-bd01-00144feabdc0.html#axzz1YPbjkRoj

22 http://www.guardian.co.uk/business/2011/jun/06/willie-walsh-emissions-trading-scheme

23 *EU ETS and Aviation*, p. 1

24 EU Commission MEMO/06/506, Questions & Answers on aviation & climate change, December 2006

25 *EU ETS and Aviation*, pp. 5-6

26 *EU ETS and Aviation*, pp. 7

27 World Bank, *State and Trends of the Carbon Market 2011*, June 2011, p. 9

28 http://www.nytimes.com/cwire/2011/01/10/10climatewire-china-moves-toward-carbon-emissions-trading-81996.html?pagewanted=2

29 http://www.reuters.com/article/2011/02/07/us-carbon-korea-idUSTRE7161II20110207

30 http://www.grist.org/climate-policy/2011-06-27-court-rules-californias-cap-and-trade-program-can-advance

31 *State and Trends*, p. 18

Chapter 5

1 *Improving the emissions trading scheme*, p. 8

2 http://europa.eu/rapid/pressReleasesAction.do?reference=IP/07/1614&format=HTML&language=EN

3 Helm, D. & Hepburn, C., (eds), *The Economics and Politics of Climate Change*, Oxford University Press, 2009, pp. 242-43

4 'Green growth: opportunities, challenges and costs', pp. 6-7

5 'Green growth: opportunities, challenges and costs', p. 2

Chapter 6

1 http://www.parliament.uk/business/committees/committees-a-z/commons-select/energy-and-climate-change-committee/news/new-inquiry-eu-ets/

2 http://www.carbonretirement.com/sites/default/files/Government%20EUA%20revenue%20report_0.pdf

3 Prins G., *et al, The Hartwell Paper: A new direction for climate policy after the crash of 2009*, May 2010, p. 11

4 PriceWaterhouseCoopers, *Carbon Taxes vs Carbon Trading: Pros, cons and the case for a hybrid approach*, March 2009, p. 15

5 Committee on Climate Change, *Meeting Carbon Budgets-the need for a step change*, October 2009, p. 68

6 HM Treasury, *Carbon price floor: support and certainty for low-carbon investment*, December 2010, 2.8

7 Energy and Climate Change Select Committee, *Carbon Price Support*, 6.138

8 *The role of carbon markets*, 4.66

9 *Chain Reactions*, p. 14

10 *Chain Reactions*, p. 2

11 Energy and Climate Change Select Committee, *Carbon Price Support*, 6.145

12 Uncorrected oral evidence taken before the Energy and Climate Change Committee, 4 March 2009

13 DECC, *Carbon price floor: support and certainty for low-carbon investment*, December 2010, p. 23

14 Energy and Climate Change Select Committee, *Carbon Price Support*, 6.177-78

15 http://www.euractiv.com/en/energy-efficiency/brussels-disarray-energy-directive-news-505654

16 http://uk.reuters.com/article/2011/06/14/us-eu-energy-climate-idUKTRE75D4FO20110614

17 International Air Transport Association, *Full Support to Accelerate Single European Sky*, April 2010

18 Transport Committee, *The Use of Airspace – European Developments*, July 2009, 5.73

19 http://www.margueritefund.eu, Fund Rationale page

20 http://www.bis.gov.uk/news/topstories/2011/May/green-investment-bank

21 'To create 48kWh per day of offshore wind per person in the UK would require 60 million tons of concrete and steel...to make 48kWh of nuclear power person in the UK would require 8 million tons of steel and 0.14 million tons of concrete.' MacKay, D. *Sustainable Energy – Without the Hot Air*, Cambridge, 2008, p. 62

Chapter 7

1 *Green growth: opportunities, challenges and costs*, p. 14

2 *Carbon Budgets*, p. 4

3 DECC, *UK Greenhouse Gas Emissions 1990-2010 (provisional), headline results*, Table 1

4 European Commission, *EU Energy in Figures 2010, CO_2 emissions by sector*, Tables 1.1 & 1.21

5 McKinsey & Company, *Making the Green Journey Work, Optimised pathways to reach 2050 abatement targets with lower costs and improved feasibility*, February 2011, p. 10

6 MacKay, O., *Sustainable Energy – Without the Hot Air*, Cambridge: UIT Cambridge, September 2010, p.114

7 DECC, *Statutory Security of Supply Report*, November 2010, p. 6

8 Constable, J., *The Green Mirage: why a low-carbon economy may be further off than we think*, London: Civitas, July 2011, p.xiv

9 *The role of carbon markets*, 4.84

10 *Green growth: opportunities, challenges and costs*, p. 3

11 Ipsos Mori poll, 18 April 2011

12 *Memorandum submitted by David Newbery*, Paragraph 21

13 Mott MacDonald, *UK Electricity Generation Costs Update*, June 2010, p. 69, Figure 7.6

14 *Green growth: opportunities, challenges and costs*, p. 13

15 http://www.bellona.org/news/news_2011/Porto_Tolle_CCS

16 *Sustainable Energy*, p. 221

17 Merlin-Jones, D., *The Industrial and Commercial Finance Corporation: Lessons from the past for the future*, October 2010, available: http://www.civitas.org.uk/pdf/ICFC.pdf

18 *Carbon Budgets*, p. 12

19 *Carbon Budgets*, p. 11

20 Netherlands Environmental Assessment Agency, *No growth in total global emissions in 2009*, June 2011

21 *Statutory Security of Supply Report*, p. 11

22 https://restats.decc.gov.uk/cms/national-renewables-statistics/

23 http://www.ft.com/cms/s/0/c49ce7d2-6da5-11de-8b19-00144feabdc0.html#ixzz1SpbEWd2C

24 Figures based on a load factor of 0.3 and each wind turbine producing 1MW. Extra renewable capacity required by 2020 is 200,989,660 MWh to 80,607,530 MWh according to different DECC scenarios

Chapter 8

1 *Carbon Budgets*, p. 17

2 *Green growth: opportunities, challenges and costs*, p. 7

3 Committee on Climate Change, First Report of the Committee, *Building a low-carbon economy – The UK's contribution to tackling climate change*, December 2008, p.106

4 *The Green Mirage*, p. 105

5 http://www.decc.gov.uk/assets/decc/statistics/projections/71-uk-electricity-generation-costs-update-.pdf, Figure 7.1

6 Recalculated from six billion population in *Sustainable Energy*, p.164

7 http://www.ossfoundation.us/projects/energy/nuclear and http://www.columbia.edu/~jeh1/mailings/2009/20090713_Strategies.pdf

8 For further information, this is a good place to start: http://energyfromthorium.com/about/

9 Turan Ünak, 'Thorium: bright start of the next nuclear era, *International Nuclear Chemistry Society News*, Vol 6. No. 1 (January 2009), p. 20

10 National Nuclear Laboratory, *The Thorium Fuel Cycle: An independent assessment* (August 2010), p. 2

11 *Sustainable Energy*, p. 168

12 http://www.bbc.co.uk/news/uk-england-merseyside-13875032

13 http://www.bbc.co.uk/news/uk-england-somerset-11564284

14 http://www.bbc.co.uk/news/uk-scotland-scotland-business-13865843

15 http://www.checkmateuk.com/seaenergy/

16 http://www.desertec.org/en/global-mission/focus-region-eu-mena/

17 http://www.reuters.com/article/2009/08/24/us-energy-maghreb-desertec-sb-idUSTRE57N01720090824?sp=true

Chapter 9

1 *Chain Reactions*, p. 38

2 *The role of carbon markets*, 4.74

3 *Economist*, 'Doffing the cap', June 14th 2007

4 *Carbon Taxes vs Carbon Trading*, p. 17

5 Green Investment Bank Commission, *Unlocking investment to deliver Britain's low carbon future*, June 2010, p. 1

Chapter 10

1 http://www.decc.gov.uk/en/content/cms/meeting_energy/nuclear/nuclear.aspx

2 International Energy Agency, CO_2 emissions per kWh of electricity and heat generation, *CO_2 emissions from fuel combustion highlights*, (2010 edition) p. 107

3 *Sustainable Energy*, p. 169

4 *Statutory Security of Supply Report*, p. 9 Figure available here: http://www.inference.phy.cam.ac.uk/withouthotair/c27/page_211.shtml

5 DECC, *Government response to Dr Mike Weightman's interim report*, June 2011

6 Sizewell B's output is approximately 1.2GW and DECC estimates 12GW will be taken offline by the IED

7 *Sustainable Energy*, p. 171

8 Allison, W., *Radiation and Reason: The Impact of Science on a Culture of Fear*, October 2009, p. 198

9 *Sustainable Energy*, p. 211

10 http://business.scotsman.com/business/Gas-is-answer-to-UK39s.6716653.jp

11 McKinsey & Company, *Making the Green Journey Work, Optimised pathways to reach 2050 abatement targets with lower costs and improved feasibility*, February 2011, p. 24

12 *Making the Green Journey Work*, p. 25

13 http://www.statistics.gov.uk/cci/nugget.asp?id=155

14 DECC, *UK Greenhouse Gas Emissions 1990-2010 (provisional)*, Table 1, CO_2 emissions: 1990: 589.7 $MtCO_2e$ 2010: 491.7 This is a reduction of 16.6 per cent so far. Plus 5.8 per cent is 22.4 per cent.

15 *Making the Green Journey Work*, p. 29

16 http://www.bgs.ac.uk/research/energy/shaleGas.html and http://www.guardian.co.uk/business/2011/mar/01/fracking-shale-gas-energy-mps

17 Ofgem, *Project Discovery Energy Market Scenarios*, 122/09, October 2009, p. 5

18 http://www.sussex.ac.uk/newsandevents/pressrelease/id/3220

19 http://www.sussex.ac.uk/newsandevents/pressrelease/id/3220

Conclusion

1 Europa, 'Emissions Trading EU Environment Commissioner Margot Wallström welcomes Council agreement as landmark decision for combating climate change', December 2002, IP/02/1832

2 Prins, G., *et al*, *The Hartwell Paper*, p. 5

Appendices

1 Bluestein, J., *Upstream Regulation of CO_2*, Presentation to the National Commission on Energy Policy Workshop, Washington DC, 2005.

2 Ofgem, *Updated Household Energy Bills explained*, January 2011 and EU ETS figure of £13 given on request from Ofgem

3 http://www.eprg.group.cam.ac.uk/wp-content/uploads/2008/11/eprg0617.pdf

4 Assumes current median domestic user consumers 16.5MWh at a cost of £36.85/MWh and that existing environmental policies represent £1.33/MWh of that. Also assumes a £30 tax would add £5.52/MWh to the price.

5 DECC, *Estimated impacts of energy and climate change policies on energy prices and bills*, July 2010, p.6

6 DECC, Provisional estimates of the impacts of energy and climate change policies on energy prices and bills of large energy intensive users. http://www.decc.gov.uk/assets/decc/11/about-us/economics-social-research/2350-pro-ests-of-impacts-energy-and-prices.pdf

7 DECC, *The UK Fuel Poverty Strategy: 7th Annual Progress Report 2009*, October 2009, p.5

8 DECC, *Fuel Poverty Statistics 2011*, July 2011

9 http://www.decc.gov.uk/assets/decc/statistics/fuelpoverty/1_20091021091505_e_@@_ukfuelpovertystrategy7annreport09.pdf, p2.

10 http://www.decc.gov.uk/en/content/cms/news/csr_hmt_releas/csr_hmt_releas.aspx

11 Warm Front, *Connecting with communities: the Warm Front Scheme Annual Report 2009/10*, p. 4

12 http://www.nao.org.uk/publications/0809/the_warm_front_scheme.aspx

13 http://www.official-documents.gov.uk/document/other/9780108508172/9780108508172.pdf

[14] http://www.nao.org.uk/publications/0809/the_warm_front_scheme.aspx

[15] http://www.decc.gov.uk/assets/decc/What%20we%20do/Supporting%20consumers/Addressing%20fuel%20poverty/1281-warm-front-annual-report-2009-10.pdf , using total expenditure, households helped and estimated annual saving.

[16] http://www.decc.gov.uk/assets/decc/Statistics/fuelpoverty/2183-fuel-poverty-2009-detailed-tables.pdf Table 7.

[17] http://www.ofgem.gov.uk/Sustainability/Environment/RenewablObl/Documents1/RO%20Costs%202011_12%20Consultation.pdf

[18] *The Green Mirage*, p. 105

[19] http://www.ofgem.gov.uk/Sustainability/Environment/fits/Newsletter/Documents1/Feed-in%20Tariff%20%28FIT%29%20Update%20Newsletter%20Issue%204.pdf

[20] http://www.energysavingtrust.org.uk/Generate-your-own-energy/Cashback-Calculator

[21] http://ec.europa.eu/clima/policies/ets/linking/index_en.htm

[22] Directive 2OO9/29/EC – Article 11a(8). Available from Page 78, http://eur-lex.europa.eu/LexUriServ/LexUriServ.do?uri=OJ:L:2009:140:0063:0087:en:PDF

[23] http://www.decc.gov.uk/assets/decc/statistics/projections/71-uk-electricity-generation-costs-update-.pdf

[24] E.g. Oldbury site summary, Lifetime plan 2006-7, available at http://www.nda.gov.uk/documents/loader.cfm?csModule=security/getfile&PageID=4002

[25] http://www.independent.co.uk/news/business/news/true-price-of-uks-nuclear-legacy-163160bn-472368.html

[26] http://www.decc.gov.uk/assets/decc/11/stats/publications/dukes/2306-dukes-2011-chapter-4-natural-gas.pdf

[27] http://www.decc.gov.uk/assets/decc/11/stats/publications/dukes/2304-dukes-2011-chapter-2.pdf

[28] http://dataservice.eea.europa.eu/PivotApp/pivot.aspx?pivotid=475